R. D. MILLS
50 STAPLOE
ST. NEOTS
CAMBS.

XMAS 78.

A colour guide to familiar
MAMMALS

A colour guide to familiar

MAMMALS

by Dr Vladimír Hanák
Illustrated by Květoslav Hísek

OCTOPUS BOOKS

Translated by Dr Stanislava Pošustová
Graphic design: Soňa Valoušková

English version first published 1977 by
OCTOPUS BOOKS LIMITED
59 Grosvenor Street, London W1

© 1977 Artia, Prague

ISBN 0 7064 0609 5

Printed in Czechoslovakia
3/10/15/51-01

CONTENTS

WHAT ARE MAMMALS?

Most people can recognize a mammal when they see one and can distinguish a mammal from other classes of vertebrate animals. It is difficult, however, to choose one particular character which is true for all mammal species, because the mammals as a group are so diverse. There are some mammals which appear to resemble other groups of vertebrates more closely. Bats with their specialized wings and ability to fly could be mistaken for birds, and whales which are completely aquatic may appear to be gigantic fish. Even a study of the early history of mammals does not provide a common character for mammals. Those features characteristic of modern mammals have evolved one at a time in the evolution of the mammals from the reptiles.

What are mammals then? They comprise the most advanced and most adaptable vertebrate class. Most of the organs which function within a mammal have reached a high degree of development and have allowed mammals to adapt to many varied habitats and environments. The most important organ to undergo this development is undoubtedly the brain, especially the outer region or cortex, which results in greater mental ability. Although the brains of different mammals vary in structure and size, the brain of the mammal is relatively large in comparison with that of lower vertebrates, and it can be said that the development and increase in size of the brain are basic characteristics of the whole mammalian class.

Many people will, however, think that the production of milk and the special way that the young suckle from the mammary glands is a more typical mammalian characteristic. Certainly this feature only occurs in mammals and has become the basis for naming the whole class. The embryo spends a long time within the womb of the mother, and even after birth the young are dependent on the mother for a considerable period, during which time they are nourished by milk from the mammary glands. These glands have a common origin with sweat and sebaceous glands. They occur in pairs down the underside of the mother's body, and the number matches the number of young produced at any one time.

Another familiar and important characteristic feature of mammals is that they give birth to live young. In most mammals a special organ called the placenta develops inside the mother's body, and connects the developing foetus with the inner lining of the womb. The placenta enables the foetus to be protected and nourished by the mother so that the young are well developed at birth. This mode of reproduction has been perfected in most mammals, which are called placental mammals. In most marsupials, however, the placenta only acts as a loose, short-term connection between the embryo and the mother's body. As a result, the young are born at an embryonic stage and their further development continues while the young is attached to one of the mother's nipples, and it is usually protected by the ventral pouch or marsupium. The spiny anteater and the duck-billed platypus do not bear live young but, like their reptilian ancestors, reproduce by laying eggs.

As feathers are characteristic of birds, so fur is a characteristic of mammals. Fur is made up of hairs, which are outgrowths of the skin made of a protein called keratin. The underfur is formed of dense, short, soft hairs, between which protrude long, coarse guard hairs. Specialized hairs also occur, such as whiskers (vibrissae), which are rooted in pouches with nerve endings, and spines. Adult individuals of some species, however, do not possess fur, but their young have at some time in their embryonic development possessed hairs, and so the absence or reduction of fur is a secondary feature. Examples of mammals lacking fur are whales, hippopotamuses, and rhinoceroses. In other mammals the reduction of fur is associated with the development of an internal shell, as in armadillos, or of horny scales, as in pangolins. Similarly, the spines of a hedgehog or porcupine and the horns of a rhinoceros are nothing else but highly modified hairs.

In connection with fur, one of the most advanced adaptations of mammals must be mentioned, namely that they are warm-blooded—that is, they can maintain a constant body temperature. The majority of mammals have a body temperature of between 36 °C and 40 °C, but in marsupials the body temperature is lower, ranging from 27 °C to 32 °C. A constant high body temperature is necessary if the organs within the body are to work efficiently, and mammals can maintain such a temperature even in extreme environmental conditions. This has enabled them to colonize cooler areas of the earth.

The structure of the skull and skeleton of mammals also shows certain characteristic features. The skeleton comprises about 200 bones. Some bones which are present in other vertebrate classes have been lost in the mammals. For instance, mammals retain only a single bone in the lower jaw. In itself this may not seem important, but for palaeontologists it is often the only feature to distinguish a primitive mammal from a reptile. Mammals are the only vertebrates with three internal ear bones (hammer, anvil and stirrup) and in mammals there are almost always seven neck vertebrae.

There are four basic types of mammalian teeth: incisors, canines, pre-molars and molars. The teeth of other vertebrates are not normally distinguishable into types. The number and shape of teeth vary between different groups, and these differences form a basis of the modern classification of mammals.

Mammals have inherited from their reptilian ancestors four five-toed limbs, and the majority of mammals still retain these. In a few groups, however, the limbs have undergone structural changes involving fusion of bones, elongation of limbs and reduction in the number of digits. Ungulates serve as an example of extreme reduction of digits: in horses, for instance, only the middle (third) digit remains and this carries the whole weight of the body. The whole limb has also been straightened, resulting in horses walking on the tips of the

9

toes instead of on the whole surface of the sole. In some mammalian orders, limbs evolved to meet the requirements of living in a medium other than on land. In truly aquatic mammals, such as whales and sea cows, the forelimbs have been converted to paddles and the hindlimbs have disappeared. In sea lions, both pairs of limbs appear like fins and they have sacrificed some of their agility on land. In bats, the original five-toed limb remains, but the forearm and all four fingers of the forelimb have become elongated. The wings of a bat thus resemble more the wings of early flying reptiles than those of birds; in birds the bones are reduced, and the wing surface is formed by specialized flight feathers.

There are also some characteristics to be found in the soft, internal organs of mammals. The heart is perfectly divided into four compartments, but this feature is also developed in birds and crocodiles. Animals with a constant body temperature and a highly developed brain need to have oxygenated and deoxygenated blood separated if they are to be very active. Because of this, the blood supply to and from the lungs needs to be separated from the supply to and from the rest of the body. Unlike other vertebrates, mammalian red corpuscles do not have a nucleus. The whole corpuscle can therefore be used for carrying oxygen.

Other peculiarities of the mammal body include the epiglottis, which is a cartilaginous flap which closes the larynx while the animal is swallowing, the complex facial muscles which allow facial expression, and the diaphragm. The diaphragm is a sheet of muscle separating the chest and abdominal cavities, which makes ventilation of the lungs more effective and thus ensures a greater supply of oxygen to the blood.

When all these adaptations and characters operate together, the result is a unique type of animal—the mammal. The highly developed mammalian brain could not function if the oxygen supply was inadequate, and this is ensured by the four-chambered heart and the diaphragm. Nor could the brain function if the body temperature fluctuated, and this is maintained with the aid of an insulating layer, the fur coat.

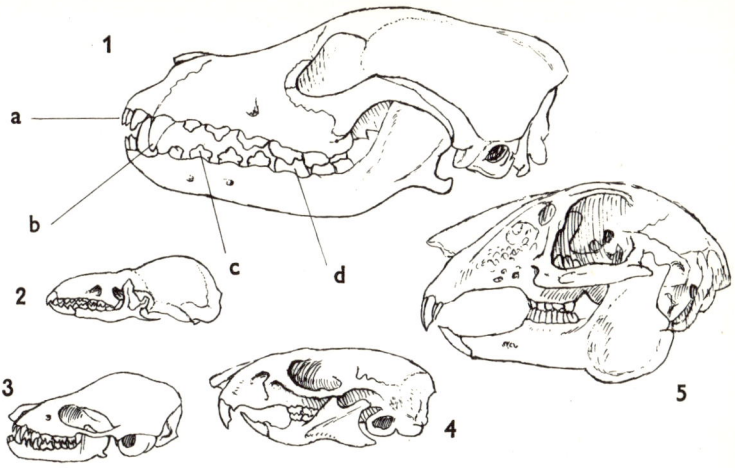

Fig. 1. Skulls of some mammals:
1 — dog; a — incisors, b — canine, c — premolars, d — molars;
2 — shrew, 3 — bat, 4 — house mouse, 5 — hare.

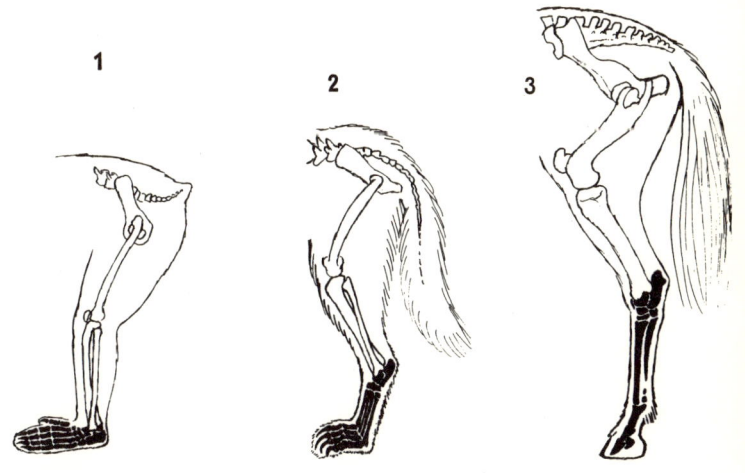

Fig. 2. Types of hind legs:
1 — plantigrade, 2 — digitigrade, 3 — unguligrade.

Overheating is also prevented by means of sweat glands, which are unique to mammals, and we have already seen how the skin glands have been converted to mammary glands fulfilling quite a different function.

Mammals are generally larger in size than other vertebrates, but there are immense differences within the class. The smallest mammal in the world is the Etruscan shrew *(Suncus etruscus)*, which weighs about 2.5 g. The largest living mammal is the blue whale *(Balaenoptera musculus)*, which can attain a length of 33 m and a weight of 130 t. The largest mammal inhabiting dry land is the African elephant which weighs up to 6 t, surpassed in size only by some extinct rhinoceros relatives.

At present about 4,500 mammal species live in the world, which is only just over 10% of all known vertebrate species. However, their importance remains indisputable. The most important domestic and laboratory animals, game animals and agricultural pests are to be found among the mammals. Despite this, mammals in the wild have not been studied as thoroughly as other animals, probably because many are nocturnal and secretive. Nowadays, however, research is increasing in many fields and interesting discoveries have recently been made about some mammal groups, for example whales, bats, rodents and primates.

THE DIVERSITY OF MAMMALS

Mammals first appeared on the Earth about 200 million years ago during the period of time which we call the Mesozoic era. They evolved from reptiles and their immediate ancestors were small forms of the mammal-like reptiles (order Therapsida). The early mammals were very inconspicuous animals and were overshadowed by the very large reptiles such as the dinosaurs. It was not until the end of the Mesozoic era, some 70 million years ago, that they became a really important part of Nature. The following brief survey is intended to show only the diversity of living mammals, and to give an account of the characters which distinguish the orders of placental mammals which have representatives in Europe and North America.

Insectivores (order INSECTIVORA) are regarded as the most primitive placental mammals. They exhibit a number of characteristic features which are considered primitive from an evolutionary point of view. The body of an insectivore is small and has four five-toed limbs, and the whole surface of the sole of the foot bears the weight of the body when moving. Insectivores have between 24 and 38 teeth, but the dentition varies between the different families. The brain is relatively small and smooth, with none of the complex foldings or convolutions found in the brains of higher mammals. The internal ear is not very well protected by bone as in higher mammals.

The most important sensory organ of an insectivore is its nose. It has a long tapering snout, and associated with this keen sense of smell are large lobes within the brain called olfactory lobes. These guide the animal within its environment, using the information received through the snout. Insectivores do not form a uniform group and they are usually classified into several distinct sub-orders, of which only one occurs in Europe.

The three most widespread families of insectivores are the hedgehogs (Erinaceidae), moles (Talpidae) and shrews (Soricidae). In their habits, shrews probably show the closest resemblance to the ancestral insectivores. There are about 270 species throughout the world, with the exception of Australia and the greater part of South America where insectivores as a whole are absent.

Moles are adapted to an underground or aquatic life, and about 20 species inhabit the temperate zone of Eurasia and North America. Hedgehogs also have special features, among them the spiny coat and the system of muscles just beneath the skin which enables them to curl up into a ball. Several species of hedgehog inhabit the Old World.

Insectivores live mainly on animal food such as insects and other invertebrates, small vertebrates, birds' eggs and carrion. Only occasionally will they eat vegetable matter. Their bodies work at a very high rate, and so small species cannot tolerate long fasts. They are therefore continually searching for food.

About 370 species of insectivores are known and in some areas of the world they are so numerous that they form an important component of the mammalian fauna.

Bats (order CHIROPTERA) form the second most numerous order of mammals, and about 1,000 species occur throughout the world. They are very widely distributed, and they are the only mammals on some oceanic islands. Towards the poles, however, their distribution is limited by temperature, but a few species do inhabit regions near the Arctic Circle.

Structurally, bats reveal a close relationship with the insectivores. The teeth of a bat, like those of insectivores, are numerous and sharp. The forelimbs are elongated to form the wing skeleton, which supports a bare, transparent, doubled skin, the flying membrane or patagium. Another conspicuous feature of many bats is that they have a small finger-like projection just in front of the earhole which stands up like a smaller, second ear flap opposite the main one. This is

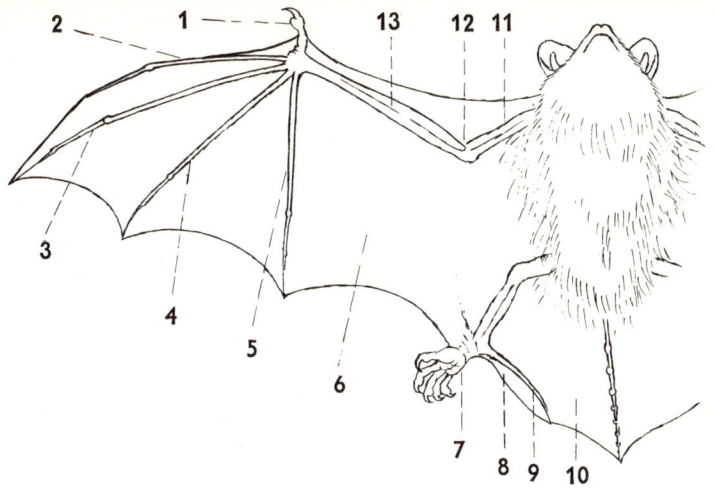

Fig. 3. Bat's wing:
1 — first digit (thumb), 2 — second digit, 3 — third digit, 4 — fourth digit,
5 — fifth digit, 6 — wing membrane, 7 — heel, 8 — post-calcarial lobe,
9 — calcar, 10 — tail membrane, 11 — shoulder, 12 — elbow, 13 — forearm.

called the tragus and can take many forms, varying from species to species. The majority of bats have developed a way of orientating themselves, based on the principle of echolocation. They give out a series of very high frequency sounds which are inaudible to human ears, and perceive the echoes which are reflected by obstacles. The tragus is thought to focus the stream of echoes received. In their way, bats are able to distinguish the shape and distance of an object even in complete darkness. They use this system to locate their food, which consists of flying insects.

Bats can be classified into two sub-orders, Megachiroptera and Microchiroptera. As the name implies, the Megachiroptera includes mostly large forms with a wingspan of up to 1.7 m. The majority of these bats are fruit-eating, and they live in the Tropics. The remaining 17 families of smaller, and mostly insectivorous, forms belong to the sub-order Microchiroptera. The commonest families are the horseshoe bats (Rhinolophidae) and evening bats (Vespertilionidae). There are about 68 species of horseshoe bat; they are easily recognized by the curious membranous outgrowth on the snout, the nose leaf, and by the absence of a tragus. The majority of European bats, however, belong to the family of evening bats, of which there are about 300 species in the world. The members of this family feed solely on insects, which they catch on the wing at dusk and during the night. They spend the daytime resting in dark shelters, for example in caves, attics, mines or hollow trees.

The order of **primates** (order PRIMATES) was so named as this is the order to which we ourselves belong. There are about 195 living representatives which live mainly in tropical regions.

Although many primates, especially Man and apes, can be said in some respects to have reached a peak of evolutionary development, their physical structure still retains some primitive features. In this respect they resemble the modern insectivores. In fact, it is impossible to separate some groups of insectivores from the most primitive primates. The tree shrews (Tupaidae), for example, are sometimes classified with the primates and sometimes with the insectivores as they possess both primitive insectivore features as well as specialized primate ones. They can be considered as living examples of the stage in evolution when the primates were just emerging from within the original insectivores.

All primates possess high mental faculties, and this is what makes them basically different from other mammals. Their limbs have long, flexible digits which have nails instead of claws. The thumb can be moved so as to oppose the other fingers, and this enables them to clasp branches when they are moving about in trees. In Man, this is an important feature in the development of the human hand. The majority of primates live in trees, but some have adapted their way of life to grasslands and rocky areas. Many primates are diurnal in habit and are sexually active all the year round. Females usually give birth to one offspring at a time, and the young are dependent on the mother for a relatively long period of their development. All these characteristic features can be seen in the evolution of the most advanced primate — Man.

There is hardly a place in the world where **rodents** (order RODENTIA) are completely absent. They are the most numerous of all mammals both in the number of species and the density of distribution. Most rodents are small or medium-sized animals which live on vegetable matter, although some are omnivorous. Their characteristic feature is that they have only two incisor teeth in each jaw, and these teeth never stop growing. These teeth wear each other down by their opposing actions, and each tooth assumes the shape of a chisel. Rodents do not possess canine teeth, and the resulting gap between the cheek teeth and incisors is called the diastema. The molars, too, have a characteristic structure: they are wide, with blunt projections in mice and hamsters, and with a ridged gnawing surface in voles.

Rodents are capable of gnawing into the hardest food materials, and this is accomplished by the use of powerful jaw muscles. They are also well known for having an immense reproductive capacity which, in favourable conditions, causes over-production of some species, making them pests to fields, forests and food stocks.

Rodents are the most diverse mammalian order. They can be divided into 7 sub-orders and 33 families. Those common to Europe and North America are the squirrels (Sciuridae), beavers (Castoridae), hamsters and New World rats and mice (Cricetidae), voles (Microtidae), and birch and jumping mice (Zapodidae). Other families which are found in Europe are the mole rats (Spalacidae) and the Old World rats and mice (Muridae). North American families include the pocket gophers (Geomyidae), which show similar adaptations to the European mole rats, and pocket mice and kangaroo rats (Heteromyidae). The American kangaroo rats show remarkable resemblances to the Old World rats (Dipodidae). The Americas have also been colonized by three members of the Old World family Muridae, namely the house mouse, the black rat and the brown rat. One particularly important group of rodents (sub-order Caviomorpha) is almost exclusively found in South America. They include the guinea pig, capybara, coypu, chinchilla and the North American porcupine.

Rodents are a very successful group biologically. There are about 2,000 species and some groups are still evolving, as shown by the spread of black rats and brown rats and the evolution of new strains of house mice in isolated areas.

There are about 240 species of **beasts of prey** (order CARNIVORA) which can be found throughout the world, with the exception of Australia whose only representative, the dingo, was brought to that continent by its original inhabitants. Carnivores have several common characteristic features associated with their body structure and mode of life. They have teeth specially adapted for eating flesh. Each half of their jaw contains three incisors and one greatly enlarged canine tooth which is used for seizing and killing prey. The last upper premolar and the first lower molar have been transformed into shearing teeth with sharp-edged crowns, which function in a scissor-like way. Carnivores have a massive skull with a large crest running along the top of the brain case, to which powerful jaw muscles are attached. Their brain is comparatively large, as a result of having well-developed senses of smell, hearing and sight, and a high degree of intelligence. To hunt successfully, one needs mobility, strength, intelligence and keen senses, and carnivores possess all of these.

The cats (Felidae) possess an acute sense of smell and sight and are capable of moving lightly and quietly. With the exception of the cheetah, they all possess retractile claws. They catch their prey by stealth, either by stalking or by ambush.

Members of the weasel family (Mustelidae) are smaller carnivores with an elongated body and short limbs. They all have well-developed scent glands at the base of the tail. The secretions of this gland are strong and are used for marking the territory. In skunks, these scent glands have been modified for defence. Different mustelids hunt their prey in different ways.

Weasels and stoats hunt mostly on the ground, occasionally following their prey — small mammals and sometimes rabbits — into their burrows. Martens find their prey, which consists mainly of squirrels, in trees. Otters feed on fish and occasionally shellfish. Many mustelids have valuable fur which in the past has led to their persecution and, in some regions, to their extinction.

Members of the dog family (Canidae) have a relatively long snout with large jaws and between 38 and 48 teeth. Larger forms, such as the wolf, hunt in packs and run their prey down. They have a keen sense of smell, and their hearing is acute.

Bears (Ursidae) are the largest carnivores, and they have a worldwide distribution. They have thick-set bodies and very short tails. When they walk, the whole surface of the sole touches the ground (plantigrade), and this is one of the most typical characteristics of bears. Bears in general are the most omnivorous of the carnivores. Their sense of smell is well developed.

Civets and their relatives (Viverridae) are slender, long-necked, short-legged carnivores which mostly inhabit Africa and the warm parts of Asia, but are also found in south-west Europe. They are regarded as the most primitive carnivores, and although they resemble cats in their appearance and movement, they are more closely related to hyaenas. Like the Mustelidae, the family is diverse and includes, besides the civets, the genets and the mongooses.

The raccoon family (Procyonidae) are an American family of carnivores, but one species, the raccoon *(Procyon lotor)* is now found in Europe, having escaped from captivity. The family also includes the pandas of Asia.

Rabbits, hares and pikas, or **lagomorphs,** (order LAGOMOR-PHA) are vegetarian, and they possess a number of features in common with rodents. However, they differ from the latter with respect to some features of the skull and teeth. Lagomorphs have another pair of small incisors behind the large ones in the upper jaw and, unlike most rodents, they are unable to seize food with their front paws. They can, however, utilize otherwise indigestible plant material: the swallowed food is partly digested and enriched with protein and vitamin B with the help of bacteria in a specialized part of the intestine. The resulting soft, pulpy pellets are eaten again and passed through the intestine a second time. The food is therefore fully made use of, and the final indigestible remains form the familiar hard pellets.

Lagomorphs are divided into two families: hares and rabbits (Leporidae) and pikas (Ochotonidae). Hares and rabbits are distributed throughout the whole world, including Australia and New Zealand where they were introduced by Man. Pikas are small lagomorphs, found in the Rocky Mountains of North America and mountainous regions of Asia.

The hooved mammals, or **ungulates,** can be divided into those which have an even number of toes on each foot (order ARTIODACTYLA) and those with an odd number of toes (order PERISSODACTYLA). The characteristic feature of an ungulate is that the tip of each toe is surrounded by a horny hoof, and the animal walks on this hoof and thus on the tips of its toes. The whole leg is long and slim and is well adapted for running. In most even-toed ungulates, only two digits are fully developed, the third and fourth. The second and fifth digits, almost as large in the hippopotamus, are moderate in size in pigs and deer, and reduced or absent in giraffes and cattle. The majority of even-toed ungulates are vegetarian; only the pigs are in any way omnivorous.

The more primitive even-toed ungulates (non-ruminants) comprise the pigs, peccaries and the hippopotamuses. The pigs (Suidae) are found in most of the Old World, the peccaries (Tayassuidae) are found in various parts of the New World, and the hippopotamus is now confined to Africa.

The ruminants (camels, deer and cattle), like the rabbits, can utilize indigestible plant material. They possess a four-chambered stomach. The grazed food is accumulated in the spacious first chamber (rumen) where bacteria act upon it. The semi-digested solid material, the cud, is then regurgitated into the mouth, chewed for the second time during the animal's rest and, only then, completely digested. Deer (Cervidae) are also ruminants, but they are characterized by their antlers, which are bony structures which grow on the frontal bone of the skull and are shed every year. The most numerous, diverse family of ruminants are the horned ruminants (Bovidae). Their horns are permanent structures formed by horny sheaths covering outgrowths of the frontal bone. This family includes several species of cattle, goats, sheep and gazelles.

There are about 115 species of bovids living today. Bovids have spread widely over the grassy steppes of North America and Asia and the group is still evolving. The ancestors of many of our domestic animals are found within this group.

The American antelope, or pronghorn, belongs to a closely related family, the Antilocapridae, which differs in that the horns are shed annually.

The hey-day of the odd-toed ungulates (order PERISSODAC-TYLA), when there were many forms related to our modern rhinoceroses and tapirs, is already past. However, domesticated representatives of this order, the horse and ass, have played an important role in the history of Man. They are characterized mainly by the structure of the leg. The single third digit is alone retained and carries the whole weight of the body. The remaining digits are always smaller or rudimentary, as in the horse. The gradual reduction of the digits, and the transformation of the original five-toed foot into the present one-toed foot, can be traced from the remote ancestors to the present day in an almost unbroken record. Today, only 20 species of odd-toed ungulates exist throughout the world.

The **whales and dolphins** (order CETACEA) are the mammals most perfectly adapted to their aquatic life. Their fore limbs have been converted to paddles, and the hind limbs have dissappeared. The brain is highly developed, and dolphins have been found to be very intelligent, and a great deal of scientific research is being carried out on these animals.

21

REPRODUCTION OF MAMMALS

It has already been mentioned in the opening chapter that almost all mammals bear live young. The embryo develops inside the body of the mother and is attached via the placenta to the inner lining of the womb. Oxygen and soluble food passes from the circulatory system of the mother, through the placenta to the foetus. Because of this, mammalian eggs are minute and possess very little yolk.

Reproduction in mammals is complex and varies from species to species, but it is always dependent on the influence of the sex hormones, the secretions of which are under the influence of the pituitary gland. This organ is found on the floor of the brain and is influenced by differences in the duration of daylight during the year. The majority of mammals are not sexually active during the whole year, but only at certain times. These periods of sexual activity, when sex cells — ovum and spermatozoon — mature and are able to fuse, are called oestrous or heat in the female and rut in the male. Many mammals have oestrous only once a year, and are therefore called monoestrous species. Examples are roe deer, red deer, and many beasts of prey. On the contrary, rodents, shrews and rabbits may pass through several oestrous periods in a year, and they are sexually active throughout the spring and summer and even in winter if conditions are favourable. These are called polyoestrous species, and are renowned for their great reproductive capacity.

The mating season is an important time in the life of mammals. Many mammals live alone throughout the greater part of the year, and only in the mating season do both sexes associate. The red deer stag, for instance, lives solitarily until the beginning of the rut when he collects a group of hinds and protects them fiercely. In polyoestrous species, both sexes meet more often and usually also inhabit the same territory, but they actually live together only during the short mating seasons. Males usually take no part in the rearing of the young. However, examples do exist of mammals which remain as a pair for one or more years. Among these are beavers, jackals, and some antelopes.

The mating period varies in different species of mammals, but the period is constant for each individual species. Roe deer rut in July and August, red deer in September and October, fallow deer in October and November, and wild boar as late as November to January. The majority of mammals, however, mate in the spring so that the young are born during the late spring and summer.

The time from when the egg is fertilized until the young are born is called the gestation period or pregnancy. Pregnancy is usually constant within each species, and the length of the pregnancy will depend on the size of the animal and the stage of development of the young at birth. In general, the smallest mammals, for instance small rodents and insectivores, have the shortest pregnancy. Marsupials have a very short pregnancy, in some species as short as 8 days, and, because of this, marsupials give birth to poorly developed young. Among placental mammals, the shortest pregnancy occurs in some small hamsters (16 days), and mice, voles and shrews have pregnancies lasting from only 18 to 26 days. However, even species which are closely related and are of similar size may have pregnancies of different duration. The wild rabbit, for example, gives birth to blind, naked young after only 30 days, while the pregnancy of the hare lasts up to 40 days and the young are born with fur and their eyes open.

Some mammals have an unusually long pregnancy which is not related to either their size or to the degree of development of the young. In these cases, pregnancy is referred to as latent or prolonged pregnancy. The fertilized egg undergoes only the first few divisions, after which the development of the embryo stops for a time, often for several months. The development of the young embryo is not completed until conditions are more favourable, which is usually in the spring. In female roe deer, for example, there is a long, latent period between the fertilization of the egg in July or August and the birth of the young in May or June. The existence of latent pregnancy has also been demonstrated in martens, badgers, stoats, some marsupials and armadillos. Bats in temperate regions appear to have a disproportionately long pregnancy, as they mate at the end of the summer and the young are not born until May or June of the following year. This, however, is a different phenomenon, called delayed fertilization, where the spermatozoon survives the hibernation period within the female sexual organs and not until the following spring does it fertilize the ovum which has just matured. The actual pregnancy lasts from the fertilization of the egg to the birth of the young, i. e. 53—75 days, depending on the species and the temperature of the environment. These two adaptations have evolved so that both the mating of the adults and the birth of the young can take place at the most favourable time of the year.

The number of young in a litter differs with individual species. As a rule, smaller species have larger litters, while large or advanced mammals have few young. Animals bearing large litters are some small rodents, carnivores, insectivores, oppposums and pigs. A litter of 32 young has been recorded in the tenrec, an insectivore of Madagascar. Large mammals produce one or two young, as do the majority of bats. Their low yearly reproductive rate is compensated by a long lifespan. The average number of young in an individual's litter is an inherited character, but may vary according to different climatic conditions, sources of food and density of population. If conditions are especially favourable, there may be a rapid, albeit temporary, increase in the population density, causing cycles of overpopulation in animals such as voles, lemmings and hares.

It has already been said that different mammals give birth to young at different stages of development, and this in turn depends on the way of life of the species. Ungulates of the open grassland give birth to their young in an unprotected terrain; consequently they need to be well developed at birth and after a very short time are able to follow their mothers.

Newborn hares are also well developed, and born with fur. They have to fend for themselves from the very beginning of their life and are born in an open nest without the protection of a burrow. Young carnivores are usually born in burrows and are born with fur, but are blind. Young bears are particularly small and poorly developed at birth; the newborn of the brown bear, for instance, are the size of a rat. Whales, dolphins and sea lions all give birth to well-developed young — newborn sea lions and seals are half the length of the adult animals. The majority of small mammals, such as rodents, insectivores and bats have poorly developed young which are born naked, blind and weak. They are reared carefully by the mother and protected in burrows or shelters. An interesting adaptation occurs in young bats, which spend the first weeks of their life together with their mothers on the rafters of attics or on cave walls. The young are born with limbs adapted for climbing and a very well-developed thumb. They are also capable of holding on to objects with their milk teeth, and can therefore move quickly about over the walls and cling to the mother's body from the first few days of their life. Young bats remain in colonies until their first independent flight, and they are only transferred to another place if danger threatens. This is also true of other mammals, especially rodents, carnivores and some insectivores.

Many mammals, for example primates, although well developed at birth, need a comparatively long time for further development. Others which are poorly developed at birth have a more rapid development. Examples of these mammals are small rodents, insectivores and bats. The newborn mouse grows fur and opens its eyes within the first week of its life and soon becomes independent of the mother.

Small rodents attain sexual maturity in the first few weeks of their life, while shrews are not sexually mature until the following spring. Bats usually become sexually mature in the second year of life. In larger mammals, sexual maturity is attained much later. Red deer, for example, are sexually mature after 2 to 4 years, apes after the tenth year and elephants not until they are between 12 and 20 years old.

MIGRATION

Most mammals occupy the same territory all their life and never leave the boundaries of a certain area which is known as their home range. However, the hunting areas of some species, particularly beasts of prey, may be fairly extensive and so their hunting expeditions may cover many scores of kilometres. Regular seasonal migrations resembling those of birds are exceptional. Migrations are usually motivated by the search for new sources of food, more favourable climatic conditions, suitable shelters or, sometimes, by the need to find a mate. Northern reindeer populations, for example, travel every autumn in large herds to the south where there is less snow and where it is possible to find food, returning again in the spring via the same routes. Herds of African ungulates, for example gnus, zebras and antelopes, make their way across the grasslands in cycles corresponding with the seasons of the year.

Sea lions regularly migrate each year from the open sea to certain islands in order to mate. They gather in large numbers in their regular meeting places where the females give birth to the previous year's young, after which mating takes place again. The mass migration of lemmings, small rodents inhabiting the more northerly regions, occurs as a result of their cyclic overpopulation. These journeys are only made by the surplus of the lemming population from their homeland, and the numbers of migrating animals are gradually reduced by the rough travelling conditions and by birds and beasts of prey, which usually gather in large numbers along the lemmings' migration routes.

Bats of the temperate zone also migrate regularly, but their journeys from the summer habitats to the wintering grounds are motivated by the need to find suitable places for hibernation, rather than by the lack of food. The flights of most species are relatively short, being no more than 100—200 kilometres. Some species of bats, however, undertake regular long-distance flights which may be compared with the migration of birds. The northern populations of the noctule bat probably cover distances of several thousands of kilometres in order to hibernate in the mild climate of the Caucasian foothills, the Crimea, the Balkans and southern France.

Long-distance migrations are also undertaken by some whales. These whales feed on plankton and, as this varies geographically according to the temperature of the sea water, they have to travel long distances to keep up with their food supply. Each species has its own regular routes and places to which all the animals return. Many species of large whales gather in the icy waters of the Antarctic Ocean during the short summer months, and towards the autumn they all move northwards as far as the tropical seas. In three months, they have to cover a distance of as much as 10,000 kilometres. Experienced whalers are usually well acquainted with the whales' sea routes and take advantage of this knowledge.

Regular migrations over shorter distances are known in some mountain-dwelling mammals, for example the chamois and the red deer. When winter approaches, they descend to the wooded foothill regions and mountain valleys where more food and more favourable climatic conditions can be found. Even some small rodents, which are generally considered to be resident animals, become more active towards the autumn. When days become colder, woodmice and even small shrews start to inhabit houses, particularly isolated ones. Populations of house mice and rats are known to leave urban areas for the summer, some of them returning for the winter.

Usually, any increase in a mammal's range is a result of migration. Species which today remain exclusively within one particular territory originally had to colonize their present habitats, some members of the species having moved into it from elsewhere. This type of migration is called emigration and there are recent examples of this among mammals. The muskrat, for instance, was introduced artificially into central Bohemia, but within the last 10 years it has colonized most of central Europe. The British Isles have been similarly colonized by the American grey squirrel, and nowadays we are witnessing the rapid spread of the coypu which was introduced from South America as a fur animal and has colonized large areas of Britain, Europe and North America. With increasing numbers of conservation areas and national parks, the range and density of many rare species has begun to increase.

MAMMALS IN WINTER

One of the most interesting seasonal features of some mammals is their ability to survive the unfavourable weather conditions during the winter by hibernating. This is a state of torpor which is very similar to normal sleep, but during which the activity of the body organs is slowed down to a minimum to reduce the consumption of energy and thus lessen or eliminate the need to eat. Consequently, hibernation in its purest form has evolved in mammals inhabiting the cool-temperate zone, particularly in those species which feed on insects and the green parts of plants, which are not found in the wild during the winter. Examples of animals which hibernate are hedgehogs, bats, ground squirrels, marmots, hamsters, dormice and jumping mice. All mammals, however, possess the tendency to slow down their metabolic rate when subjected to extreme cold. This is particularly so in the more primitive mammals such as the egg-laying species, many marsupials, armadillos and sloths. Many other mammals of the temperate zone are also known to pass through a period called false hibernation, during which the body temperature, the heart rate and the metabolic rate are all only slightly lowered. Such false hibernations occur in bears, badgers and raccoons. These animals wake up several times during the winter, sometimes for quite a long time, when they may also eat. Female bears even give birth to their young at this time. Squirrels and field mice are able to pass through short spells of severe weather in a lethargic state.

The main factors influencing the onset of hibernation are the autumnal drop in temperature, the shortening day length, the increasing lack of food and the increase in the fat reserves within the animal's body due to autumn over-feeding. The carbon dioxide content of the ill-ventilated burrow also increases, and is a contributory factor to the onset of hibernation. These changes are received by the brain, and the glands which produce the appropriate hormones are stimulated.

Most mammals which hibernate spend the winter in underground burrows lined with dry plant material. They remain completely still, and rest in their characteristic sleeping position. Dormice, for example, are curled up into a small furry ball, the head lying against the tail. Bats hibernate in cracks and crevices of caves or hang freely suspended from ceiling or walls. Horseshoe bats enclose all their body within the soft flying membrane. The body of a hibernating animal feels cold to the touch and its breathing and heart beat are slow. In ground squirrels, for example, the respiration rate is reduced from 150—200 breaths per minute to 1—4 per minute. In bats, there are 3-minute periods when the animal breathes 25—50 times per minute, and these short periods alternate with 3—5-minute pauses when the bat does not take a breath at all. In rodents, the heart rate falls from 200 beats per minute to 5—18 beats per minute. Hibernating mammals do not react to any environmental stimuli, with the exception of temperature, which indicates that the activity of the brain is also suppressed. It is hard to believe that such a state of absolute torpor can last for as long as seven months out of twelve. Indeed, scientific experiments have shown that all hibernating animals do wake up from time to time for a short, or even longer, period during which they may take food from their stocks, or stretch their limbs a little before returning to sleep. Hedgehogs and hamsters are known to wake quite often, bats less frequently, and ground squirrels, dormice, and marmots hardly at all.

In certain steppe animals, especially ground squirrels, jerboas, gerbils and birch mice, a phenomenon occurs called summer sleep or aestivation. Physiologically, this is identical with winter sleep and it helps the animals to survive periods of sultry weather and droughts. Sometimes aestivation may pass directly into hibernation, so that the animal spends most of the year in a lethargic state.

PLATES

Hedgehog
Erinaceus europaeus

Erinaceidae

Hedgehogs are well-known and very popular mammals. Most interesting is their peculiar coat consisting of several thousands of tough prickly spines, which are in fact modified hair. This unusual body cover and the ability to curl up instantly into an invulnerable spiny ball provide the hedgehog with an ingenious protection against all enemies. It leaves its shelter after dusk in search for food, being easily identified on its hunting expeditions by its noisy stamping, accompanied by loud snuffling. It consumes large amounts of food, mainly insects, worms, molluscs and small vertebrates. Its diet also includes venomous snakes which exhaust themselves against the hedgehog's spines. It is not, however, completely immune from snake venom. Hedgehogs are also very fond of birds' eggs and young and therefore are not very welcome visitors to pheasantries. In spite of this, they are generally considered useful as they destroy large amounts of harmful insects. In most European countries hedgehogs are protected by law, but many of them die every year under the wheels of cars and many others are poisoned by various chemicals used in pest control. The hedgehog spends the winter sleeping in a burrow thickly lined with dry plant material. It leads a solitary life and confines its activities to a relatively small home range. It can be found over much of Europe and Asia; in Europe it occurs almost everywhere, with the exception of the northernmost parts and some islands, from the lowlands up to the tree line. There are two distinct colour forms of hedgehog in Europe, a Western brown-bellied form and an Eastern white-bellied form.

Body length:
200 — 290 mm.
Tail length:
20 — 45 mm.
Weight:
700 — 1,200 g.
Litter:
3 — 10 young once or twice a year.
Life span:
8 — 10 years.

Head of the Eastern white-bellied form

Common Shrew

Sorex araneus

Of all the European shrews, the common shrew is the most abundant. It can be found everywhere in regions with enough dampness and some undergrowth for it to hide in, from the lowlands high up into the mountains, and is most abundant in woods, shrub growths and meadows with dense, tall grass. It is a tiny, active animal with a pointed snout and a coat which is at first brownish but later turns blackish-brown. It feeds on insects, spiders, small molluscs and worms and can be met with during the day as well as at night, both in summer and in winter. It is a very voracious creature, consuming daily its own body weight of food, which it chews quickly with its 32 sharp teeth with brown-red tips. The reason for its voracity is quite simply explained: it needs lots of energy for its constant activity and, being a small animal, it has a relatively large surface area over which to lose body heat. Shrews only rarely dig their own burrows, usually making use of those of other small mammals, but sometimes they build nests of dry leaves in tree stumps or between roots, under stones or in long grass. The female gives birth to the young from April to September. They are born blind and hairless but grow very quickly, attaining their parents' size and appearance within some three weeks. The common shrew inhabits all of Europe, with the exception of Ireland and the northernmost regions. In Asia it is widespread north of the steppe belt as far as Japan. In America, members of the genus *Sorex* are widespread, and are generally called long-tailed shrews. In some years their numbers increase considerably.

Body length:
65 — 85 mm.
Tail length:
32 — 47 mm.
Weight:
5 — 12 g.
Litter:
5 — 10 young,
three or four
times a year.
Life span:
approx. 16 — 18
months.

Pygmy Shrew
Sorex minutus

The pygmy shrew is one of the smallest mammals in the world. It differs from its closest relatives, particularly the common shrew, by its smaller dimensions, by its brownish coat which does not change colour in adult animals, and mainly by its relatively long, bushier tail which seems slightly depressed at the base. Its habitats are similar to those of the common shrew, but it is found in even denser vegetation. In spite of its small size, it seems to be more resilient against cold and dampness than the common shrew and can therefore be found even on moorlands at higher altitudes. The habitats of these two related species often overlap without any apparent enmity between the animals, and if they meet they never attack one another. The pygmy shrew is one of the most widespread and abundant mammals in Europe, even though it is five to ten times less abundant than the common shrew. Like all other shrews, it is very vulnerable to starvation and will die quickly if its food supply runs out. For this reason it is very difficult to keep in a terrarium. Very little is known about its way of life, but it seems to differ only very slightly from that of its relatives. Its area of distribution is roughly the same as that of the common shrew, except that it has penetrated to Ireland and some northern European islands. In southern Europe, however, its occurrence is limited almost exclusively to hilly and mountainous regions. In the damp coniferous forests of Finland and in the north of the USSR, one may come across its close relative, *Sorex minutissimus* (illustrated below), which is even smaller.

Body length:
45 — 60 mm.
Tail length:
32 — 46 mm.
Weight:
3 — 5 g.
Litter:
5 — 8 young once or twice a year.
Life span:
16 — 18 months.

Alpine Shrew

Sorex alpinus

This shrew is only found in the mountainous regions of central and southern Europe, from the Pyrenees as far as the Carpathians and the Balkans. In the past, however, it was quite abundant throughout the whole of Europe, including lowland areas. It occurs most frequently in the dense weed growths near mountain streams, on mountain meadows and in scree fields, where it always seeks the dampest and most shaded spots. Along streams and torrents, it descends in certain places even to the foothills, down to altitudes of 300 — 400 m, but it is most at home between the altitudes of 700 — 2,000 m. The Alpine shrew can be very easily identified: its coat is greyish black and its tail is strikingly long and white on the underside. It is, however, sometimes mistaken for the water shrew, from which it differs only in its hairless, light paws, which are flesh red in living animals. Its way of life is the least known of all the European shrews. It probably feeds mainly on mountain molluscs, but certainly also on other invertebrates. The breeding season is from April to mid-September. As in all other shrews, the young of the Alpine shrew grow very quickly and are probably sexually mature in the first year of their life.

Body length:
62 — 77 mm.
Tail length:
54 — 75 mm.
Weight:
6 — 10 g.
Litter:
5 — 7 young twice to three times a year.
Life span:
16 — 18 months.

Water Shrew

Neomys fodiens

Of all the European shrews, only the water shrew is a good swimmer. It is also a skilled diver and can even run underwater. Surprisingly, its anatomy is only slightly adapted for this unusual way of life: the coat is thick and waterproof, the wide hind paws are bordered with a rim of lengthened tough bristles which act like a web, and the underside of its tail is equipped with a keel-like ridge of stiff hairs. It is a marvellous experience to watch this animal hunting in water. It is active both day and night and, when diving, it resembles a moving silver ball, the silver sheen being due to the reflection of light on the air bubbles which remain attached to its fur. It moves swiftly on the water surface as well, with its snout projecting upwards, and it can run without difficulty on the river bed. It is equally agile out of water, on dry land. The greater part of the water shrew's diet (mainly aquatic insects and their larvae, freshwater shrimps, aquatic molluscs and fish fry) is found in the water. Sometimes it does considerable damage to fish fry, and is therefore not very popular with fishermen. It is particularly fond of streams with densely overgrown banks, which provide ample hiding places among roots and stones, and sometimes inhabits the overgrown parts of still waters. Its range comprises the forest belt of Asia as far as the Far East and the whole of Europe, except Ireland and the Mediterranean region. Its relative, the Mediterranean water shrew *(Neomys anomalus)*, which is not so closely associated with water, also occurs in Europe.

Body length:
70 — 96 mm.
Tail length:
47 — 77 mm.
Weight:
10 — 20 g.
Litter:
4 — 11 young twice to four times a year.
Life span:
about 1 — 2 years.

Lesser White-toothed Shrew or Scilly Shrew

Crocidura suaveolens

Unlike the other European and American shrews, white-toothed shrews have completely white teeth, without the characteristic red tips. Their tails are also different — besides the normal hairy cover, they are thinly interspersed with long hairs. White-toothed shrews were originally inhabitants of grasslands, and consequently they are found in dryer places than other shrews. In central Europe they now inhabit mostly open spaces. During the winter they usually move into haystacks, stony walls and even houses. Due to their great adaptability and resilience, they have even penetrated forest and high-mountain regions. The most abundant in central Europe is the lesser white-toothed shrew, which can be readily identified by its smaller size and by the gradual transition of the greyish coloration of its sides into a lighter shade on the underside. It is more abundant in the warmer European regions as far north as central Germany and central France, also occurring in Asia and North Africa. In Britain, it is found only on Scilly and Jersey. The female leads her half-grown young in Indian file: the young walk in line one behind the other after their mother, each holding with its teeth the hairs at the base of the tail of the preceding animal. Europe is also the home of the common European white-toothed shrew *(Crocidura russula)* (bottom), which is larger and darker in colour, and the bicolour white-toothed shrew *(Crocidura leucodon)* (centre), whose grey-coloured sides are distinctly separated from the nearly white belly. Neither of these species is found in Britain, but the common white-toothed shrew occurs on Guernsey.

Body length:
53 — 83 mm.
Tail length:
25 — 44 mm.
Weight:
3 — 7 g.
Litter:
2 — 6 young twice to four times a year.
Life span:
1 — 2 years.

Etruscan or Savi's Pygmy Shrew

Suncus etruscus

Soricidae

The Etruscan shrew, a tiny animal weighing no more than 2 g, is usually cited as the smallest mammal in the world, although it shares this distinction with the pygmy shrew as well as with several exotic bat species. It has a light-coloured grey-brown coat and a long tail which, as in all other white-toothed shrews, has long, soft hairs as well as the normal short hair cover. Also very conspicuous are its relatively large ears, which stand out distinctly from the coat. Its range of distribution covers the warm Mediterranean region—Spain, southern France, Italy and the northern parts of the Balkans—stretching also to North Africa, Crimea, Transcaucasia and Central Asia. Related forms inhabit Africa north of the Sahara, and southern Asia. It usually occurs in dry localities: overgrown screes, shrub growths and fields. It can often be found in the vicinity of human dwellings, particularly in the ruins of buildings, in stone walls and even in houses. It leads an inconspicuous life and is nowhere very abundant. Its presence in a region is often revealed only by the remnants of the animals' skulls in owl-pellets. Consequently, we know, as yet, very little about its way of life. Its diet consists mainly of small insects and spiders.

Body length:
36—52 mm.
Tail length:
24—29 mm.
Weight:
1.5—2.5 g.
Litter:
2—5 young five or six times a year.
Life span:
unknown.

Mole

Talpa europaea

The mole is excellently adapted for its under-ground existence. It has a cylindrical body cov-ered with a short, thick coat and powerful spade-shaped forelimbs with back-turned palms. There are no external ears, but the opening is protected by a narrow ridge of skin. The eyes are small and are sometimes completely covered with soft skin. The mole spends nearly all its life in its complex network of underground burrows and passages, only rarely emerging to the surface. Although classed with the insectivores, its diet consists not only of insects and their larvae, but also of earth-worms and sometimes even small vertebrates, particularly frogs. It gathers its food during the continual expeditions along its underground passageways, where it sometimes makes 'stores' of earthworms, first paralyzing them by biting into the head end. The importance and useful-ness of the mole is still a subject of dispute. It kills large numbers of harmful insects but at the same time it is itself harmful as it kills earth-worms and undermines plant roots, not forget-ting the unwelcome mole-hills it leaves in mead-ows and gardens. It can be found both in the lowlands and high up in the mountains, usually in meadows, pastures, and also in deciduous forests. It is widespread throughout Europe, with the exception of the northernmost parts, extend-ing eastwards as far as the Urals. Several similar species occur in eastern Asia and in the Mediter-ranean region and their relatives are also found in North America. The most peculiar among them is the star-nosed mole *(Condylura cristata),* which has a rosette of 22 hairless fleshy out-growths arranged round its nostrils.

Body length:
125 — 160 mm.
Tail length:
23 — 28 mm.
Weight:
65 — 120 g.
Litter:
4 — 5 young
once a year.
Life span:
3 — 4 years.

Pyrenean Desman

Galemys pyrenaicus

The Pyrenean desman, an insectivorous relative of the mole, is well adapted for life in water. It has a glossy, close-packed waterproof coat, the toes of its large hind paws are connected by a short web and the long, sparsely haired tail is compressed from side to side so that it will serve as a kind of rudder. Worth noting also is its snout, which is extended into a long, hairless, mobile trunk, with which the animal can easily seek and examine its food. The musk gland, located at the base of the tail, produces an offensive-smelling secretion. It lives in the Pyrenees and in the mountains of northern Spain and Portugal, usually in areas around rapid-flowing, clean and well-aerated mountain streams or in marshy meadows at the altitudes from 300 — 1,200 m above sea level. It digs burrows in soft banks and, like the water shrew, it gathers its food — mainly worms, molluscs, insects and small invertebrates — in the water. Its sexual activity starts as early as the end of winter, usually in January, and the young are born from March to July. In eastern Europe, particularly the Don and Volga river valleys, there lives its larger relative, the Russian desman *(Desmana moschata)*, which has been successfully introduced to some other regions of the USSR. The fur of this species is being commercially exploited. The present-day populations of these two species are no more than remnants of the great numbers of desmans which inhabited Europe in the past.

Body length:
110 — 135 mm.
Tail length:
130 — 155 mm.
Weight:
50 — 80 g.
Litter:
about 4 young
once a year.
Life span:
2 — 3 years (?).

Lesser Horseshoe Bat

Rhinolophus hipposideros

The lesser horseshoe bat is the commonest European representative of the family Rhinolophidae, which differ from common bats (Vespertilionidae) by the special membranous outgrowths on the snout, by the typical pointed shape of their ears, the lack of tragus, the relatively short tail membrane and also by certain slight peculiarities in the anatomy of the skull and in the formation of their teeth. All horseshoe bats are warmth-loving mammals and are most common of the warm regions of Africa and Asia. In the temperate zone, they are most abundant in the Mediterranean. The lesser horseshoe bat is widespread throughout nearly all of Europe and as far north as central England, western Ireland, central Germany, Poland and southern Russia. It is also found in north Africa and central Asia. It is not, however, equally abundant in all these regions, and particularly towards the northern boundary of its range it frequents only sheltered localities with enough underground hiding places for hibernation. Originally the lesser horseshoe bat was a cave-dweller; today it is found in lofts, in belfries and in old, abandoned houses. It sleeps throughout the winter in warm mining galleries, in caves and cellars. During hibernation it hangs itself head downwards from the ceiling, covering its whole body with its soft wing membranes, so that it looks like a suspended pear. It lives in large colonies both in winter and in summer. It leaves its hiding place to hunt only after dusk, and flies close to the ground catching small insects.

Body length:
37 — 43 mm.
Forearm length:
34 — 42 mm.
Wingspan:
190 — 225 mm.
Weight:
3.5 — 10 g.
Litter:
1 young
once a year.
Life span:
up to 18 years.

Mouse-eared Bat

Myotis myotis

Vespertilionidae

The mouse-eared bat is one of the largest and also one of the commonest bats of Continental Europe. Like all other members of this genus, it has translucent membranous ears which are pointed at the ends. Its mouth is equipped with 38 small teeth. A warmth-loving, originally cave-dwelling species, it is most abundant in southern and central Europe and, apart from a single colony, is completely absent in the British Isles and in Scandinavia. It occurs also in North Africa, in the Caucasus, Asia Minor and western Asia. Further east, in tropical Asia, the very similar Eastern bat occurs, which is also widespread in southern Europe. In its central European habitats, the mouse-eared bat usually spends the summer hidden in the loft of an old house, where the females and their young form large colonies of 50 to 500 individuals. It hibernates in various caves, mining galleries and large cellars, either singly or in groups. It can cover relatively long distances (as much as 260 km) flying from its summer habitats to the wintering grounds. As a rule, it remains faithful for all its life to a single summer roost and a single winter roost. The hairless and blind young are born from the end of May to the beginning of June in the summer habitats. After a month, when they have reached the size of their parents, they are already able to fly independently. The mouse-eared bat starts hunting only after dusk, when it visits housing estates, parks, gardens, avenues of trees and the margins of woods. Its diet consists mostly of larger insects, particularly moths, and also large beetles.

Body length:
60 — 80 mm.
Forearm length:
55 — 68 mm.
Wingspan:
350 — 430 mm.
Weight:
20 — 40 g.
Litter:
1 young
once a year.
Life span:
up to 15 years.

Water or Daubenton's Bat
Myotis daubentonii
Whiskered Bat
Myotis mystacinus

Both these bats belong to the genus *Myotis.* Specialists are able to distinguish them from each other by certain minor differences in their size, the shape of the ears, and the method of attachment of the flying membrane to the feet. The water bat, which is one of the commoner species, has relatively large feet and the membrane is attached to the sole. It is found over much of Eurasia as far as 64°N, from Portugal and the British Isles as far as east Japan. It usually occurs in areas with abundant still or flowing water. During the summer, the females and their young gather in medium-large colonies in tree cavities, wall crevices, tunnels and under bridges. This bat will sometimes travel over long distances (up to 100 km) to find a suitable place for hibernation — usually a cave or a mining gallery, where it stays either suspended freely from the ceiling or hidden in a crevice. It starts hunting after dusk and often remains on the wing all the night, usually flying close to the water surface.

The whiskered bat is another common species. It is slightly smaller, and the flying membrane is attached to its toes. It is widely distributed over the whole of Eurasia up to the Arctic Circle, and also in north Africa. It is found in all types of habitats from the lowlands up to the mountains and it also frequents water habitats. During the summer months it hides in tree cavities and in abandoned houses. In winter individual bats hibernate in cellars, mining galleries and caves.

In North America there are numerous species of the genus *Myotis,* of which the little brown bat *(Myotis lucifugus)* is the commonest and most widespread.

Myotis daubentonii
Body length:
40 — 48 mm.
Forearm length:
33 — 39 mm.
Wingspan:
210 — 250 mm.
Weight:
6.5 — 10 g.

Myotis mystacinus:
Body length:
35 — 48 mm.
Forearm length:
31.5 — 36 mm.
Weight:
4.5 — 6.5 g.
Litter:
both species 1
young once
a year.
Life span:
approx. 5, max.
about 15 years.

Noctule Bat
Nyctalus noctula
Pipistrelle or Common Bat
Pipistrellus pipistrellus

Vespertilionidae

The noctule bat is a representative of the so-called tree bats, i.e. those which spend the summer period almost exclusively in the cavities of old trees. The noctule is moderately common over a wide area comprising Eurasia, except the northernmost parts, and north-western Africa. Its favoured habitats—deciduous and mixed forests, parks and avenues of trees are usually found in the lowlands and plateaux. During the summer, the females form colonies of some 20 to 60 individuals in tree cavities, whereas the males lead a solitary life. In winter, both males and females hibernate in large groups consisting of several hundreds or thousands of bats. Their most common hibernating sites are tree cavities. The northern populations in particular have to cover very long distances (1,000—2,000 km) when flying to their winter quarters. Noctule bats set about their hunting expeditions soon after sunset, at first hunting at a moderate height in a zigzag flight and only later, with the growing darkness, descending closer to the earth.

The pipistrelle is the smallest European bat. Its range comprises nearly the whole of Europe as far north as 60°N, the greater part of Asia and north Africa, and it is very common in all these areas. During its autumn migration flight it often visits buildings in large numbers. In summer it hides behind window shutters, in wall and tree crevices and in birds' nest-boxes. It spends the winter hidden deep in the crevices of cave walls, in mining galleries, in cellars and often also behind the frames of pictures in old houses. It never starts hunting before dusk. Bats similar to the Eurasian pipistrelles occur over much of the United States.

Nyctalus noctula:
Body length:
60—80 mm.
Forearm length:
47—56 mm.
Wingspan:
300—400 mm.
Weight:
15—40 g.
Litter:
2 young
once a year.
Life span:
up to 8 years.

Pipistrellus pipistrellus:
Body length:
33—45 mm.
Forearm length:
27—34 mm.
Wingspan:
180—210 mm.
Weight:
3.5—8 g.
Litter:
2 young once a year.
Life span:
up to 8 years.

Head of
Pipistrellus pipistrellus.

Long-eared Bat
Plecotus auritus
Barbastelle
Barbastella barbastellus

Vespertilionidae

The long-eared bat, as its name suggests, has extremely long ears, in fact the longest of all European species. Surprisingly enough, it has only recently emerged that two very similar species exist in the Eurasian region. These two species are very difficult to distinguish from each other. The long-eared bat is the hardier species and is therefore also more common towards the north (as far as 63°N). In central Europe it inhabits mainly tree-covered hilly regions, and in more southerly areas it occurs only at higher elevations.

Both these species share the same habitats in Europe and Asia as far as China and Japan. In summer they live in colonies in lofts, birds' nest-boxes and tree cavities. From October to the end of March they hibernate individually in caves, mining galleries and cellars, but also in various buildings, where they usually hide in crevices, although they sometimes perch on open walls. When they are hibernating, their long ears are hidden under the flying membranes, so that only the narrow tragus is visible, like a minature ear. They start hunting at dusk, flying slowly and skilfully round the crowns of trees, and sometimes even gathering insects from the leaves.

The North American big-eared bats *(Plecotus rafinesquei* and *Plecotus macrotis)* are found over much of the United States and Mexico, and are similar in habits.

The barbastelle is a small bat with a short, slightly flattened snout. Its blackish-brown coat is a lighter shade on the back. It spends the summer in lofts or wall crevices and hibernates in galleries and caves, sometimes forming large colonies comprising hundreds of individuals.

Plecotus auritus:
Body length:
42—51 mm.
Forearm length:
35—42 mm.
Wingspan:
220—265 mm.
Weight:
4.8—8.2 g.
Litter:
1 young
once a year.
Life span:
12—15 years.

Barbastella barbastellus:
Body length:
47—52 mm.
Forearm length:
38—41 mm.
Wingspan:
250—275 mm.
Weight:
6—10 g.
Litter:
1 young
once a year.
Life span:
15 (max. 20) years.

Pine Marten
Martes martes
Beech Marten
Martes foina

Martens are some of the best-known representatives of the mustelid family, which are slender, short-legged, small beasts of prey. They are very cautious, wary animals which set out for their hunting expeditions only towards the evening or at dawn. Their excellent sight and hearing is well known. Both species are widespread over nearly the whole of temperate Eurasia.

The pine marten, which is a strict tree-dweller, is an adept climber. It can even jump skilfully from tree to tree. It makes its nest in tree cavities and abandoned squirrels' nests. Trees are also its favourite hunting grounds, as it preys mainly on squirrels which it pursues along the branches. It also feeds on small rodents and birds, sometimes supplementing its diet with eggs and fruit. Its hunting grounds, a part of a forest where it chooses several hiding places, is marked out by secretions from the stink glands which are located at the base of its tail. The pine marten's reproduction is also of interest, particularly the so-called latent pregnancy.

The beech marten used to occur in remote rocky locations; today, however, it is mostly found in the vicinity of Man—near human dwellings and sometimes even in large cities. It hides in rock and wall crevices, in lofts, etc., and usually hunts its prey on the ground. Its diet mostly consists of rats, but sometimes the beech marten will also prey on poultry.

The American marten *(Martes americana)* and the fisher *(Martes pennanti)* of the northern American forests are similar in habits and appearance.

Martes martes:
Body length:
480 — 530 mm.
Tail length:
230 — 280 mm.
Weight:
1.2 — 1.6 kg.
Litter:
3 — 5 young
once a year.
Life span:
8 — 10 years.

Martes foina:
Body length:
450 — 500 mm.
Tail length:
250 — 270 mm.
Weight:
1.7 — 2.1 kg.
Litter:
2 — 6 young
once a year.
Life span:
up to 15 years.

Weasel
Mustela nivalis
Stoat
Mustela erminea

The weasel and the stoat, close relatives, are also, with their small, slender bodies, typical representatives of the family Mustelidae. They differ from each other in size and particularly in the colour of the tail tip, which in the stoat is always black both in summer and winter. Furthermore, stoats develop a white winter coat, whereas this never happens to weasels, with the exception of the northern populations. The weasel, the smallest of all European mustelids, inhabits open country with fields and meadows, also forest margins and uncultivated shrubbery, and is moderately abundant in all these habitats. It feeds mainly on small rodents, which, thanks to its small dimensions, it is even able to pursue into their own burrows. Damage caused by the weasel to game is generally quite negligible. There are well-marked size differences between the sexes in both the weasel and the stoat, the females being noticeably smaller. In addition, however, there are also individuals which are so much smaller than the rest of the population that they were until recently regarded as a different species. The weasel's range of occurrence includes the whole of Europe except Spain, Ireland, Corsica and Iceland. It also lives in North Africa and all the temperate belt of Asia as far as Japan. The stoat's European area of distribution is limited in the south by the Pyrenees and the Alps, and it also occurs in Ireland and Greenland, the whole of Siberia and North America (where it is known as the short-tail weasel). The phenomenon of latent pregnancy has also been observed in the stoat.

Mustela nivalis:
Body length:
130—240 mm.
Tail length:
50—70 mm.
Weight:
75—130 g.
Litter:
5—7 young once (twice) a year.
Life span:
approx. 1—2, max. 5—8 years.

Mustela erminea:
Body length:
240—290 mm.
Tail length:
80—90 mm.
Weight:
150—260 g.
Litter:
4—9 young once a year.
Life span:
approx. 2, max. 7—10 years.

European Polecat
Putorius putorius
Asiatic Polecat
Putorius eversmanni

Polecats have a stink gland located below the base of their tails, the secretion from which they use both to mark out their territory and also to scare off potential enemies. The indigenous species in Europe is the European polecat, which occurs in lowland as well as mountain areas, and in various types of country — fields, shrubland, gardens and even built-up areas. Being a good swimmer and diver, it is particularly fond of water habitats, where it catches frogs and fishes. Its main diet, however, consists of small rodents, birds and their eggs, insects, worms and molluscs. It will also attack rabbits, muskrats and pheasants. It is particularly hated by farmers because of its raids on hen-houses. The polecat's hiding place is usually a well-sheltered corner in a barn or stable, or a pile of stones, and it will also construct its own underground burrow. It hunts mainly after dusk.

The Asiatic polecat's original home was in the vast steppe regions of the East, but it has spread westwards with the expansion of agriculture. The area of its distribution reaches today as far west as eastern Austria, Czechoslovakia and Poland. It inhabits steppes, pastures and fields, where it digs its own burrows. Its diet consists mainly of field rodents, particularly ground squirrels and hamsters, but it also hunts reptiles, amphibians and birds. The ferret, which is used by hunters to drive wild rabbits out of their burrows, is probably a domesticated form of one of these polecats. The American black-footed ferret *(Mustela nigripes),* though not closely related, is similar in many respects to these polecats.

Putorius putorius:
Body length:
400 — 440 mm.
Tail length:
130 — 190 mm.
Weight:
500 g — 1.5 kg.
Litter:
4 — 5 young
once a year.
Life span:
5 — 6 years, in captivity up to 8 — 10 years.

Putorius eversmanni:
Body length:
320 — 560 mm.
Tail length:
70 — 180 mm.
Weight:
500g — 1.3 kg.
Litter:
3 — 7 young
once a year.
Life span:
unknown.

Badger
Meles meles

Although the badger also belongs to the mustelids, it is much larger than martens and polecats, has a stout body and is plantigrade, i. e. it treads on the whole of its sole when walking. Its coat is also characteristic, being greyish on the body and white on the head, with one wide black band running across the eye on either side. Its feet are equipped with strong claws which the animal uses adeptly when digging its complicated, deep underground passages. These lead into a set, which provides both a shelter for the day and a place for hibernation. The badger does not normally leave its set until it is completely dark and even then it is usually extremely cautious. Its winter sleep is not very deep, its body temperature does not drop and it wakes up several times during the winter, sometimes even leaving the set. Of all the mustelids, the badger shows the greatest tendency to omnivorousness. It feeds on small rodents, insects, earthworms, slugs, but also on carrion, birds' eggs, seeds, berries, plant roots and mushrooms. In spite of this, it has been quite unjustly persecuted by hunters and its sets destroyed. Today, however, it is legally protected in most countries. The badger is widespread over nearly the whole of Europe and the temperate part of Asia as far as China and Japan. It is particularly fond of country with scattered copses and can be found at any altitude from lowland to high-mountain elevations. The American badger, *Taxidea taxus,* is similar in many respects, but is generally found in treeless country, especially the prairies.

Body length:
700 — 850 mm.
Tail length:
110 — 180 mm.
Weight:
7.5 — 15 (20) kg.
Litter:
1 — 5 (usually 2)
young once
a year.
Life span:
up to 15 years.

Otter
Lutra lutra

The otter differs considerably from the common mustelid type by its perfect adaptation to an amphibious way of life. It has a longish, agile body and a wide, flat-topped head with long tactile whiskers and short ears. The tail is long and strongly thickened at the base, and the toes are connected by a short web. The otter's coat, short and thick, is very much prized as fur. In the past it was quite abundant near waters throughout Europe, occurring also in north Africa and in Asia as far as Java and Sumatra, a similar species being found in North America. As a result of many years of severe persecution, when it was hunted both for its fur and as a threat to fisheries, it has completely disappeared from many areas and is very rare in others. It frequents both flowing and still waters, mostly in localities with inaccessible, overgrown banks in the upper reaches of rivers. Its present situation is uncertain, as otters are shy, wandering animals and change their habitats very quickly. In the water they move nearly as skilfully as fishes, which form, together with crayfish, frogs and dryland vertebrates, the main part of the otter's diet. Its holt, or den, is dug in high river banks, with the entrance often located under the water surface.

Body length:
650 — 800 mm.
Tail length:
350 — 500 mm.
Weight:
5.5 — 10 kg.
Litter:
2 — 4 young
once a year.
Life span:
15 — 18 years.

Wolverine or Glutton

Gulo gulo

Mustelidae

The wolverine is a peculiar beast of prey with a robust body and a long-haired, dense coat, rather like a bear. It is the largest of all mustelids, attaining a weight of 20, sometimes even 30, kg. It inhabits the extensive taiga and tundra regions of Eurasia and North America, but its European distribution is today limited to Scandinavia, northern Finland and the north of the USSR; its American range comprises most of Canada. It is nowhere very abundant, and in many places and particularly in Europe it is threatened with extinction. Except during the breeding season, the wolverine leads a solitary life. The individual animals have permanent hunting grounds, which for some males may measure as much as 1,000 sq km. Wolverines hunt during the day as well as at night in these large areas. Their role in the natural food chain is quite a positive one, as they mostly feed on carrion and usually only attack weak or crippled animals. They are not, however, very popular with hunters, because of their habit of stealing game from traps and provisions from hunters' cabins. Generally, the wolverine's diet is very varied, consisting in summer of birds and their eggs, insects and insect larvae, small rodents (particularly lemmings), berries and oil-rich seeds, and in winter of larger mammals, ungulates, carrion and various chance titbits. Its method of hunting is mostly by trailing, but it will also often ambush its prey. It is also fond of stealing prey from other animals. The wolverine's mating season begins towards the end of summer and the young are born after a period of latent pregnancy at the end of winter or the first half of spring.

Body length:
700 — 830 mm.
Tail length:
160 — 250 mm.
Weight:
10 — 20 kg.
Litter:
2 — 4 (5) young
once in two years.
Life span:
15 — 18 years.

Brown Bear

Ursus arctos

The brown bear is a native inhabitant of the once-impenetrable forests of Europe, temperate Asia and North America. In this extensive area there are a number of subspecies which differ from each other in colour and size so considerably that they were formerly considered as completely separate species. Worth noting, for instance, is the large Siberian bear *(Ursus arctos beringianus)* and the North American grizzly bear *(Ursus arctos horribilis)* of Alaska and the Rockies. The largest of all brown bears is the Kodiak bear *(Ursus arctos middendorffi)*, which inhabits Kodiak Island off the coast of Alaska. Originally, the brown bear was distributed throughout the whole of Europe, including the British Isles; today it occurs in significant numbers in Europe only in the Carpathians, Scandinavia and in the remnants of the original mountain forests of the Pyrenees, Italy and the Balkans. The diet of this large mammal is extremely varied: in spring and in the autumn it predominantly consumes vegetable food but it is also fond of carcasses, its favourite delicacy being honeycombs full of honey and young bees. It supplements this food with smaller vertebrates and insects. When the salmon migrate, brown bears descend to the rivers to fish. A few individuals specialize in hunting larger prey, particularly farm cattle and game. The brown bear spends the winter hidden in its den, but its sleep is not a very deep one; it often wakes up and its body temperature does not drop. The female even gives birth to her young and rears them during the winter months (December — January).

Body length:
170 — 250 cm.
Tail length:
6 — 14 cm.
Weight:
70 — 350 kg.
Litter:
2 — 3 young
once a year.
Life span:
30 — 35 years.

European Genet

Genetta genetta

Genets form a distinct, relatively primitive family of beasts of prey, with numerous species in Africa and the warm part of Asia. Their characteristics are relatively short legs, a narrow head poised on a long neck and a long tail. Although their spotted coat makes them reminiscent of cats, they are much more closely related to hyaenas. Besides the European genet, another member of this family, the mongoose *(Herpestes ichneumon)*, also occurs in Europe. The European genet is widespread in the grasslands and semi-desert regions of Africa as far as Palestine and Algeria. Its European area of distribution covers the Balearic Islands, part of the Iberian Peninsula and southern and western France. It is particularly fond of warm, dry shrubby terrain and rocky mountain slopes. Its lair is usually located in a rock crevice, preferably hidden by dense shrubbery. The European genet is an extremely wary animal and does not leave its shelter before dark. It can move almost noiselessly along the ground and is an expert tree-climber. It is not very particular about what it eats, most often feeding on small mammals, birds and their eggs, reptiles and larger insects. The young, which are born hairless and blind, are reared in the lair for the first three months.

Body length:
470—580 mm.
Tail length:
410—480 mm.
Weight:
1—2.2 kg.
Litter:
2—4 young once or twice a year.
Life span:
10—15 years.

Wolf

Canis lupus

Canidae

The wolf is the largest native canine beast of prey in Eurasia and North America. Like all other canine beasts of prey, it is a relentless runner and catches its prey by hunting it down. Its senses of smell and hearing are extremely well developed, but its vision is rather poor. Wolves remain in families or packs the whole year round. In winter, the packs are fairly large (10 to 15 individuals) and the animals hunt together. There are numerous exaggerated stories which describe the wolf as an extraordinarily astute, blood-thirsty and dangerous animal. However, it is not, and probably never has been, really dangerous to Man, usually preferring to avoid him in the wild. It is nevertheless a tireless hunter and is able to kill even the larger ungulates such as sheep and goats, and particularly domestic dogs. There are also known cases of wolves attacking a bear in its den and cannibalism has also been recorded among these animals. The largest part of the wolf's diet is formed by small rodents, birds, carcasses and various fruits. Originally it inhabited the extensive area stretching from Europe as far east as India and Japan, and also North America down to Mexico, occurring mainly in tundras, forests and grasslands and along the margins of deserts. It used to be very abundant in Europe, but today it occurs in reduced numbers only in the mountainous regions of Spain, Italy, Scandinavia and the Balkans. It is still moderately abundant in eastern Europe, particularly in the Carpathians. In North America its range covers most of Canada and Alaska. According to the latest research, the wolf seems to be the only ancestor of all modern breeds of domestic dog.

Body length:
1,000 — 1,600 mm.
Tail length:
350 — 500 mm.
Weight:
30 — 50 (75) kg.
Litter:
3 — 8 young once (twice) a year.
Life span:
15 — 16 years.

Fox

Vulpes vulpes

Canidae

Although the fox is quite popular with many people, it is heavily persecuted by hunters and gamekeepers and is killed by all possible methods. The fact that it is still relatively abundant in Europe serves to demonstrate its extraordinary hardiness, its cautiousness and a certain degree of intelligence. It is a very retiring animal which hunts mostly at night. During the day it usually remains hidden in dense thickets or in its burrow, which is built in dry out-of-the-way places, often in rocks, overgrown ravines or bushes. It mates in January or February, and the young are usually born in March or the end of April. They are reared in burrows, called earths, which are dug in the ground, accessible through several entrances and lined with soft hair. The earths are complicated structures, which are continually enlarged and repaired and may be used for several years. Foxes usually remain in the same hunting area all their life, and they are not very keen to make long expeditions. They live either singly or in permanent pairs. During their nocturnal hunting they capture small rodents, birds or larger invertebrates, and they will even attack a small hare, a young roe deer or some domestic animal. However, they probably eat a certain amount of vegetable food as well, particularly berries and fruit. Larger prey is taken into the burrow and stored for leaner days. The fox occurs, with a number of subspecies, over the whole of Europe, in North Africa, central and northern Asia and also in North America. The tundras of Eurasia and America are the home of the Arctic fox *(Alopex lagopus),* which has very short ears and a white coat in winter.

Body length:
640 — 760 mm.
Tail length:
350 — 440 mm.
Weight:
5 — 8.5 kg.
Litter:
3 — 8 young
once a year.
Life span:
10 — 12 years.

Jackal
Canis aureus

The Asiatic jackal resembles a small wolf in appearance and is actually its close relative. However, its haunts are found in warmer regions than those of the wolf; it occurs in the area stretching from the Balkans and southern Russia across Asia Minor and Central Asia as far as India and east Africa. It mostly keeps to regions covered with dense shrubland, but it is equally abundant in marshy river lowlands and reedbeds. A relatively shy animal, it does not go hunting before dark. Its presence is usually revealed by its characteristic howling, which can be heard from a great distance. The jackal's diet is very varied: it captures small rodents, ground-nesting birds, insects and reptiles, and will not reject a carcass or the rubbish found in the vicinity of human dwellings. When hunting in packs, it will sometimes tackle larger prey as well — game, sheep and goats. It is therefore not very welcome in places where it is abundant. Some valuable information about this animal has only recently come to light in studies of the east African form. It spends the day hidden in a burrow which it digs in the shelter of shrubs. Sometimes it will also make use of a natural rock crevice or even the abandoned den of another mammal. There the female gives birth to the hairless and blind young, usually at the end of March. Within some 3 — 4 months the young are already capable of an independent existence, and are sexually mature after the first year of their life. The North American coyote, though not too closely related, is similar to the jackal, although it is more widely distributed.

Body length:
850 — 1,050 mm.
Tail length:
200 — 240 mm.
Weight:
7 — 13 kg.
Litter:
4 — 6 young
once a year.
Life span:
12 — 14 years.

Raccoon-like Dog

Nyctereutes procyonoides

The raccoon-like dog is at home in the Amuro-Ussurian region, in eastern China and in Japan. Because of its economically valuable fur, it has been repeatedly introduced into the Caucasus, Ukraine and Byelorussia since the 1930s, and to-day it is an established member of the fauna of the western USSR. It has also spread into Scandinavia, Romania, Poland, Czechoslovakia and Germany, and it is still colonizing new areas. The raccoon-like dog can hardly be considered a welcome addition to the European mammalian fauna, however; it destroys many small mammals and also large numbers of useful birds and game birds. It also feeds on amphibians, reptiles, various invertebrates and carrion. As yet very little is known about its way of life in its new home. It is a secretive animal hiding in damp forests, overgrown riverbanks and even in reedbeds, and it is active only after dusk and at night. It digs its own burrow or uses the deserted burrow of some other mammal. During the winter its activity is considerably reduced and it may even sometimes fall into a winter sleep, similar to that of the badger.

The raccoon-like dog may sometimes be mistaken for the raccoon *(Procyon lotor)* which, however, belongs to a completely different family of beasts of prey. It is slightly smaller than the raccoon-like dog and weighs about 5—6 kg. It comes from North America, where it can be found over Mexico and most of the USA. It has established itself in Europe by escaping from fur farms, and there are now resident colonies in some places, particularly in Hesse (German Federal Republic).

Nyctereutes procyonoides:
Body length:
650—800 mm.
Tail length:
150—250 mm.
Weight:
4—10 kg.
Litter:
5—7 young once a year.
Life span:
7—11 years.

European Wild Cat

Felis silvestris

Felidae

The European wild cat and the lynxes are the only two representatives of the cat family living in Europe. A number of geographical forms of the wild cat inhabit Eurasia and Africa. The European form is widespread from the Caucasus and Asia Minor across southern and central Europe to western Europe, and as far north as Scotland and the coasts of the North and Baltic Seas. In the course of the last century, however, it was completely wiped out from many parts of central Europe, and today it is only abundant in the Carpathians. Due to its recent status as a protected species, its numbers have increased to a certain extent and it has even penetrated some new areas. Like the lynx, the wild cat was originally a forest-dweller but today it prefers warmer, dry places or coppices in open country. It hunts in the evening and at night, capturing mainly small rodents, birds, reptiles and fish. Only exceptionally, when it is feeding its young, does it catch larger prey. Wild cats mate from February to March, and the kittens are born in May in tree cavities, rock crevices or the abandoned burrows of other mammals. Except when they are rearing the young, the adults lead a solitary life, hunting in an area as large as 2 sq km. The wild cat is sometimes mistaken for stray domestic cats, but it can be distinguished by its thick, bushy, cross-wise-striped tail, the end of which looks as if it has been lopped off.

Body length:
790 — 940 mm.
Tail length:
290 — 350 mm.
Weight:
2.5 — 6.5 kg.
Litter:
2 — 6 young once or twice a year.
Life span:
10 — 15 years.

Lynx

Lynx lynx

The lynx was originally widespread over the whole forest belt of Eurasia, from England and France to Siberia, Alaska and Canada. Its present limited distribution, particularly in Europe, is the result of human pressures, mainly the clearing of forests and large-scale hunting. It is only found today in significant numbers in certain parts of the USSR, the Carpathians, eastern Poland, Scandinavia and some areas of the Balkans. A smaller form, the pardel-lynx, is found in Spain. Even there, however, it remains well hidden and is seen very rarely. It mostly keeps to woodlands, particularly mountain forests and larger shrub growths. The lynx hunts in the evening and again at the dawn, usually resting during the day and at night. Its hunting methods are either to attack its prey by jumping at it from an ambush or to prowl after it, relying on its keen sense of hearing and perfect vision. It has to defend its relatively large hunting area, which may sometimes cover several square kilometres; only rarely does it undertake journeys farther afield. The lair is usually built on the ground surface in a tree cavity, rock fissure or under a fallen tree; sometimes it will occupy an abandoned burrow. Its diet consists mainly of small mammals, birds and hares, but it may also attack a weak roe or red deer and some individuals are bold enough to attack a wild boar. Damage caused to game is quite negligible, however, even in areas where the lynx is fairly common.

Body length:
900 — 1,300 mm.
Tail length:
150 — 200 mm.
Weight:
13 — 38 kg.
Litter:
2 — 4 young
once a year.
Life span:
15 — 17 years.

Red Squirrel
Sciurus vulgaris
Grey Squirrel
Sciurus carolinensis

The red squirrel is a tree-dwelling rodent. When it is climbing a tree, it holds very fast to the bark with the help of its sharp claws. It can often be seen jumping from tree to tree or to the ground, moderating the fall by its widely spread limbs and by its long, bushy tail. Most of its time does it spend moving swiftly along the branches of trees, in quest of its favourite food — the seeds of coniferous and deciduous trees. It will also eat various fruits and mushrooms, and complements its diet from time to time with insects or even birds' eggs and young. This occasional theft from a bird's nest, together with the bad habit of nibbling young spruce shoots, are the only sins of this beautiful, agile animal. Its spherical nest, called a drey and made of twigs and leaves, is built in trees; sometimes squirrels will make use of a natural tree cavity. The red squirrel inhabits the woods and parks of Europe, with the exception of Iceland, the islands of the Mediterranean and the treeless regions of southern Ukraine. It is equally abundant throughout the forest belt of Asia up to Japan. The populations inhabiting this vast area differ considerably from each other in colour and size; even in Europe it is possible to come across a wide range of colour varieties from russet to a deep black. In some places squirrels are hunted and killed in large numbers for their fur.

The grey squirrel is equally popular in North America. Its original home was in the eastern USA, but in the years between 1876 and 1930 it was introduced into England, where it has multiplied to such an extent that its numbers have to be severely reduced by shooting.

Sciurus vulgaris:
Body length:
200 — 236 mm.
Tail length:
165 — 200 mm.
Weight:
250 — 400 g.
Litter:
3 — 8 young twice a year.
Life span:
8 — 10 (in captivity even 18) years.

Sciurus carolinensis:
Body length:
240 — 300 mm.
Tail length:
200 — 250 mm.
Weight:
340 — 750 g.
Litter:
1 — 6 young twice a year.
Life span:
approx. 6, sometimes up to 12 years.

Ground Squirrel
or European Souslik
Citellus citellus

Sciuridae

The souslik, although closely related to true squirrels, differs from them considerably. It lives in underground burrows and its body lacks all the typical squirrel features, such as the tufted ears and the long, bushy tail. Numerous related species inhabit the grasslands and lightly wooded regions of Eurasia and North America. The souslik is widespread from Asia Minor, southern Ukraine and the Balkans westwards to southern Poland and Czechoslovakia, and it is the only representative of this genus which penetrates from its eastern habitats as far west as central Europe. Sousliks' native home was the steppes and consequently they have colonized uncultivated land, dry pastures, etc., where they form large colonies in a fairly deep (up to 2 m) and complex network of underground burrows. The animal can often be seen running swiftly around the entrances to the burrows or sitting up in a 'begging' position. If danger threatens, it utters a warning whistle and disappears instantly underground. It is active during the day and feeds mostly on various seeds and the green parts of plants; it probably also eats insects. The souslik starts its winter sleep relatively early, towards the end of summer, and does not wake up before the end of March or even April. During this time it lives solely on the fat reserves under its skin; it does not make any food stores. In some years sousliks may overmultiply to such an extent that they cause considerable damage to field crops. They have, however, quite a number of natural enemies, mostly small mustelids and various birds of prey which, together with Man, are able to control their numbers.

Body length:
195—220 mm.
Tail length:
60—70 mm.
Weight:
240—340 g.
Litter:
6—8 young
once a year.
Life span:
4—5, max. 8—10 years.

Alpine Marmot
Marmota marmota

Sciuridae

This very popular inhabitant of the European mountains is native only to the Alps, the High Tatras and the Pyrenees, but it has been introduced to other mountain regions, such as the Low Tatras and the Black Forest. In the Balkans, on the other hand, it has been completely exterminated. Related species inhabit the mountains of Central Asia and also Siberia and North America. A relative of squirrels, it resembles a large ground squirrel. Its colonies are usually situated on the grassy and stony slopes in the dwarf pine zone, at altitudes of 1,300 — 2,750 m. It is active in the daytime, but is fairly cautious. When danger approaches, it emits the characteristic warning whistle and quickly hides in the dense underground system of its burrows and passages. These are sometimes as much as 10 m long and may go down to a depth of some 3 m. Only those individuals living near tourist chalets in the Alps seem to be bolder and not afraid of Man. Its worst enemy is the imperial eagle. Marmots live in very extreme climatic conditions, and it is therefore not surprising that their winter sleep lasts from the end of September late into the spring — to the end of April or even to May. They hibernate in their burrows, which are thickly lined with dry grass and hidden deep under the snow cover, living only on the fat reserves stored under their skin. During the summer they feed on green vegetation near their burrows. The young are born towards the end of May or in June. They are blind and hairless at birth and are suckled for the first month of their life. The Alpine marmot is legally protected in all areas of its distribution.

Body length:
530 — 730 mm.
Tail length:
130 — 160 mm.
Weight:
5 — 6 kg.
Litter:
2 — 6 young
once a year.
Life span:
15 — 18 years.

Flying Squirrel

Pteromys volans

Flying squirrels belong to the same family of rodents as true squirrels, and they resemble them both in appearance and in their way of life. Their fore and hind limbs, however, are connected by a fold of skin covered with soft hair. When this is pulled taut between the limbs it makes it possible for the animal to lengthen its leaps in a kind of gliding flight to a distance of some 35 m. The flying squirrel inhabits the mixed and coniferous woods of northern Europe and the whole of Siberia as far as the Far East. It hides in rock crevices and tree cavities, often using abandoned woodpeckers' nests. It feeds on the buds and seeds of trees, birch leaves and bark and on mushrooms and berries. Although it does not hibernate, it often remains hidden in its nest on frosty days, living on the food stored during the warmer parts of the year. It leads a secluded life and Man is often not aware of its presence, even when it occurs very close to his dwellings. The flying squirrel is well-concealed by its shyness and its nocturnal way of life, as well as by its inconspicuous coloration, which enables it to merge with the bark of trees, particularly birch. Nowadays the flying squirrel has become fairly rare in many places, the main reason being that it needs ancient forest growths for its existence and that it is very sensitive to Man's presence and to changes in the environment caused by his activities. Similar habitats in North America are the home of related species, the southern and northern flying squirrels *(Glaucomys volans* and *G. sabrinus)*.

Body length:
135 — 205 mm.
Tail length:
90 — 140 mm.
Weight:
110 — 170 g.
Litter:
2 — 4 young
once a year.
Life span:
6 — 8 years.

European Beaver
Castor fiber

The beaver is the largest rodent of Europe and North America. It inhabits banks of still and slow-flowing waters with dense vegetation. It was once widely distributed; but today it is abundant only in a few areas of the USSR and Canada. In Europe, the last small populations live around the mouth of the river Rhône, in Scandinavia, the German Democratic Republic (about 400 animals), in Poland and the USSR. In the rest of Europe, the beaver has been exterminated by Man who hunted it for its valuable fur, tasty meat (in the Middle Ages it was regarded as a fasting meal) and for the anal scent gland to which a healing power was ascribed. Beavers are perfectly adapted to an amphibious mode of life and can swim and dive well. They have a thick, waterproof coat, closeable nostrils, broad, webbed hind feet and a strikingly broadened, flat tail covered with epidermal scales. With their strong, sharp rodent teeth, they gnaw round the whole circumference of small tree trunks and fell them. In this way they obtain food — leaves and bark — as well as building material for their lodges and dams. Beavers' lodges are huge structures built over the water and made of branches and turf, with a chamber in the centre. Sometimes they also dig burrows in banks, with underwater exits. Beavers are active at night and are very agile the whole year round. The North American form is sometimes classified as a separate species, *Castor canadensis*.

Body length:
800—1,000 mm.
Tail length:
300—350 mm.
Weight:
21—30 kg.
Litter:
2—4 young
once a year.
Life span:
max. 17 years.

British Dormouse or Hazelmouse

Muscardinus avellanarius

Gliridae

This species is the smallest of all European dormice. It is the size of a small house mouse but its large dark eyes and the long, thickly haired tail distinguish it as a separate species. The hazelmouse is distributed over the whole of Europe with the exception of Spain, the Mediterranean islands, Denmark and northern Scandinavia; it extends east as far as the Volga River and Asia Minor. It is the only dormouse that is native to England. It inhabits mainly deciduous woods but, being hardier than other dormice, it occurs everywhere from lowlands high up to the mountains and even in the dwarf pine belt. It is most abundant in forests with dense undergrowth and overgrown clearings. The hazelmouse builds a spherical nest 6—12 cm in diameter, made of leaves and moss and located in bushes near, or even on, the ground. It also often makes use of tree cavities and birds' nesting boxes. Its favourite place for hibernating is a den under a layer of leaves. It starts its winter sleep in September or October, having eaten its fill during the summer, and wakes in April or May. It feeds on buds and flowers, berries, seeds and insects. It can easily be kept in captivity for, unlike its relatives, it is not irritable and does not bite. Like all dormice, it is a nocturnal animal and spends the daytime sleeping in its burrow. It can climb well and is quite at home in the branches of trees.

Body length:
75 — 86 mm.
Tail length:
55 — 77 mm.
Weight:
15 — 25 g.
Litter:
3 — 4 young once (occasionally twice) a year.
Life span:
3 — 5 years.

Edible or Fat Dormouse
Glis glis

Gliridae

The edible dormouse is the best-known Continental representative of the dormouse family whose distribution is limited to Europe, Africa and Asia. It resembles a small grey-coloured squirrel with its long, bushy tail and ability of climbing trees. The edible dormouse leads a secretive, nocturnal life, to which its large dark eyes bear witness. Like all other members of the dormouse family, the edible dormouse once inhabited open, warm deciduous woods. Now it can also be found in gardens, orchards and overgrown parks. Its area of distribution ranges from Spain across southern and central Europe to the Caucasus, Asia Minor and northern Iran. It has been successfully introduced into England. The edible dormouse is the most abundant of all European dormice and its way of life has been studied relatively thoroughly. It builds its nest, made of leaves and twigs, in the forks of branches, sometimes several metres above the ground; it also hides in tree cavities, rock crevices and even in attics of houses. When winter approaches, the edible dormouse looks for a protected shelter, usually in the roots or trees, in walls and rock fissures, where it spends up to 7 months (October—May) hibernating. It feeds on seeds, fruits, buds, leaves and insects, and occasionally even on young birds and birds' eggs. Being rather ill-natured, it is not easily bred in captivity. Like all dormice, it is legally protected.

Body length:
130—180 mm.
Tail length:
100—150 mm.
Weight:
70—120 g.
Litter:
4—7 young
once a year.
Life span:
approx. 3, max.
8—9 years.

Forest Dormouse
Dryomys nitedula
Garden Dormouse
Eliomys quercinus

These two species of European dormice are less common and less well-known than the preceding species. They are both more brightly coloured, and the conspicuous colouring of their faces and tails makes them two of the most attractive mammals. The garden dormouse is larger in size and has a bicoloured tail, broadened at the tip. Being a warmth-loving animal, it is most abundant on the European as well as African Mediterranean coasts, whence it extends as far as Central Asia. Its distribution in central Europe is only local, to the north up to the Baltic coast and central Finland, to the east as far as the Urals. In the south it inhabits thickets, and in central Europe it occurs in both deciduous and coniferous woods, parks, vineyards and in the vicinity of human dwellings. Its way of life does not differ very much from that of related species, except that it spends more time on the ground and eats larger amounts of insects.

The smaller forest dormouse's area of distribution extends from the mountains of central Asia and Caucasus across central Russia, the Carpathian region and the Balkans to Italy and through a narrow belt to the Swiss Alps. In central Europe it is most often found in deciduous and mixed woods, and in the south it also occurs in the Alpine mountain belt. Its nest is built of leaves and located in trees fairly high above the ground, and it also makes use of natural hollows and birds' nesting boxes.

Dryomys nitedula:
Body length:
80 — 115 mm.
Tail length:
70 — 100 mm.
Weight:
23 — 41 g.
Litter:
2 — 6 young once or twice a year.
Life span:
approx. 2, max. 6 years.

Eliomys quercinus:
Body length:
105 — 147 mm.
Tail length:
80 — 135 mm.
Weight:
60 — 140 g.
Litter:
2 — 8 young (average 4) once, occasionally twice a year.
Life span:
up to 8 years.

Common Hamster
Cricetus cricetus

The common hamster, a medium-sized, brightly coloured rodent, is native to the steppes of eastern Europe and western Asia. The extension of cultivated grassland brought it farther west, so that today it makes its home in the whole of central Europe as far as Belgium. However, its abundance fluctuates from one area to another. It occurs only in lowlands and hills up to about 500—600 m above sea level, in places where there is a fairly deep layer of rich soil. It lives mostly in fields, on pastures and in scrubland. It digs a relatively complex and deep network of underground passages, with a nesting chamber and several storerooms and exits. For the winter, the storeroom is filled with up to 15 kg of food which consists of grain, potatoes and sugar beet, which the hamster carries in its facial pouches. These stocks serve to feed the animal during its interrupted winter sleep. In summer it feeds on crops and weeds, as well as invertebrates and small vertebrates. It is active mainly at dusk and at night but, when the population is particularly abundant, often during the whole day. The hamster is an irritable animal. When in danger it adopts an imposing posture on its hind legs, and defends itself by snapping its teeth, snuffling and biting. Hamsters live in their burrows, usually solitarily, but the female stays with the young for a certain time. They are very prolific animals; in times of overpopulation they may even become a dangerous pest to crops. The related golden hamster, originally of western Asia, is often kept in laboratories and as a pet.

Body length:
200—340 mm.
Tail length:
25—65 mm.
Weight:
250—600 g.
Litter:
4—12 young twice to three times a year.
Life span:
6, in captivity up to 10 years.

Norwegian Lemming
Lemmus lemmus
Wood Lemming
Myopus schisticolor

Lemmings, which are related to voles, inhabit the northern regions — Scandinavia, Siberia and North America. They are well known for their periods of overpopulation which occur at almost regular intervals. During these periods, lemmings throw themselves *en masse* into streams and sea bays where they die. This extraordinary behaviour, however, applies only to some species. In appearance, the wood lemming greatly resembles a vole. It has a long, thick coat, blunt snout, short legs and tail, and is therefore well protected against the cold. It lives in the woods and peat bogs of northern Scandinavia and northern Russia; in Siberia it inhabits the taiga belt as far as Mongolia and the Far East. The network of its burrows spreads under a layer of soil or moss not far below the surface of the ground. It is not as prone as its relatives to overpopulation.

The larger and more brightly coloured Norwegian lemming inhabits the mountains and tundras of Scandinavia and the Kola Peninsula. It is a very agile and fearless animal with an immense reproductive capacity. The periods of overpopulation come at intervals of 3—4 years. During these periods, the scarcity of food causes long migrations which always end in the almost complete extermination of the over-multiplied population. In the north, lemmings form an important part of the diet of beasts of prey, mainly wolverines and foxes, and owls and other birds of prey. Several related species, the best known of which is the brown lemming *(Lemmus trimucronatus)*, live in Alaska, Canada and northern parts of the USA.

Lemmus lemmus:
Body length:
130 — 150 mm.
Tail length:
15 — 19 mm.
Weight:
40 — 112 g.
Litter:
4 — 10 young
three times a year.
Life span:
1.5 — 2 (3) years.

Myopus schisticolor:
Body length:
85 — 95 mm.
Tail length:
15 — 19 mm.
Weight:
20 — 32 g.
Litter:
3 — 7 young
twice a year.
Life span:
1.5 — 2
(occasionally 3)
years.

Bank Vole

Clethrionomys glareolus

The bank vole has the typical vole features, namely the short ears and a relatively short tail. The characteristic rusty coloration of its back, however, makes it at first sight appear different from voles of the genus *Microtus*. The bank vole is a typical inhabitant of woods, especially those with thick undergrowth, but it can also live on the edges of woods, in clearings, and even outside woods in bush-covered landscapes. It occurs from the lowlands high up into the mountains. Its area of distribution embraces the whole of Europe, with the exception of the south and extreme north, western Siberia and Asia Minor. In central Europe it is very abundant and, together with woodmice, is the commonest mammal of the forests. It is active both during the daytime and at night, spending most of the time on the ground, although it is also a good climber. Its nest, made of grass, moss and leaves, is situated not far below the surface of the ground, under stones, in tree stumps, or on the ground itself. Its diet consists mainly of the green parts of plants, seeds, roots and berries, and in winter it also eats bark and some insects. During occasional periods of overpopulation, bank voles can cause damage in forest growths, especially in tree nurseries. They live in family groups and reproduce from April to October. In the north of Europe and in Siberia two other similar species are known: the large-toothed red-backed vole *(Clethrionomys rufocanus)* and the northern red-backed vole *(Clethrionomys rutilus)*. The boreal red-backed vole *(C. gapperi)* inhabits the forests of North America.

Body length:
85—110 mm.
Tail length:
35—55 mm.
Weight:
16—36 g.
Litter:
3—5 young three to four times a year.
Life span:
12—15 months.

Water Vole
Arvicola amphibius

The water vole lives chiefly near water — brooks, rivers, ponds, lakes, pools and marshes — but it is often also found in fields, meadows and gardens. The immense area of its distribution embraces the northern parts of Eurasia from the tundra to the forest-steppe belt, from western Europe to central Siberia, covering lowlands as well as high mountains. It is found in various forms over almost all of Europe, being absent only in Ireland, the peninsula of Italy and in the southern Balkans. The form found in parts of France and in the Iberian peninsula is sometimes recognized as a separate species, *Arvicola sapidus*. Although the water vole's body shows no special adaptation to life in water, it swims and dives with great skill. It digs a network of burrows and a nesting chamber in banks, not far below the ground surface. Occasionally, especially near still water, it builds its nest in a tuft of sedges or other aquatic vegetation. Like all voles, it is active both during the daytime and at night. In favourable conditions it can overmultiply considerably. Its staple diet consists of the green parts of plants, supplemented by roots in winter. It is a serious pest of young fruit trees, gnawing their roots so that they wither. It is difficult to protect the trees against this harmful rodent because it pursues its underground activity unnoticed. It is often captured in mole traps.

Body length:
120 — 211 mm.
Tail length:
60 — 130 mm.
Weight:
80 — 200 g.
Litter:
4 — 5 young three to five times a year.
Life span:
2 — 4 years.

Muskrat
Ondatra zibethicus

The muskrat is a native of North America, where there are a number of forms from Alaska to Louisiana. Its occurrence in Europe dates from 1905 when it was introduced into central Bohemia, whence it spread in a short time over the whole of central Europe. Later the muskrat was also brought to western Europe, Scandinavia and the USSR. Today it is one of the most familiar and well-adapted mammals of this area. The muskrat, and especially its nests, can be found near still as well as running water. At the time of its initial spread, it was regarded as a pest because it dug in ponds and dykes, and also because it was erroneously believed to catch fish, but nowadays it is highly prized for its valuable fur. Today we know that the muskrat is essentially herbivorous; it eats various parts of aquatic plants and sometimes also aquatic crops, now and then adding to its diet mussels, crayfish and only occasionally some fish (in most cases dead ones). Its lair can be of two types: either it digs burrows with underwater entrances in river banks, or it builds lodges, large structures made of the roots and stems of aquatic plants, which it situates in reedbeds. Muskrats swim and dive well, using their large hind feet and long, laterally flattened, almost bare tail.

Body length:
300 — 400 mm.
Tail length:
190 — 250 mm.
Weight:
800 — 1,600 g.
Litter:
7 — 8 young three to four times a year.
Life span:
3 — 5 years.

Pine Vole

Pitymys subterraneus

At first sight, the pine vole is hardly distinguishable from similar species of voles. It is the smallest vole, however, and it has a fine, thick coat, small eyes and a relatively short tail. It is sporadically distributed, mainly in uncultivated areas, and consequently occurs most frequently at higher elevations, where it inhabits woods, mountain meadows and places along streams; in lowland areas it is often found only in forests with undergrowth. The area of its distribution extends from France to the Ukraine, and north as far as the Baltic coast. In the Mediterranean region it is replaced by other related and very similar species. Even expert zoologists cannot agree as to the classification of these forms. Different species of the genus *Pitymys* have been described from the high mountain ranges of central Europe (High Tatras, Alps). In contrast to other voles, the members of the genus *Pitymys* spend more of their life in underground burrows, and they look for food — mainly the green parts of plants and roots — in the immediate vicinity of the burrow's exit or even right inside the burrow. They live in colonies and are of nocturnal habit. They seem to live as rivals with common voles, and where one species occurs, the other is usually absent. The related American pine vole *(Pitymys pinetorum)* inhabits the coniferous forests of the eastern part of North America.

Body length:
77 — 105 mm.
Tail length:
26 — 40 mm.
Weight:
13 — 23 g.
Litter:
2 — 3 young three to five times a year.
Life span:
15 months, in captivity up to 2 years.

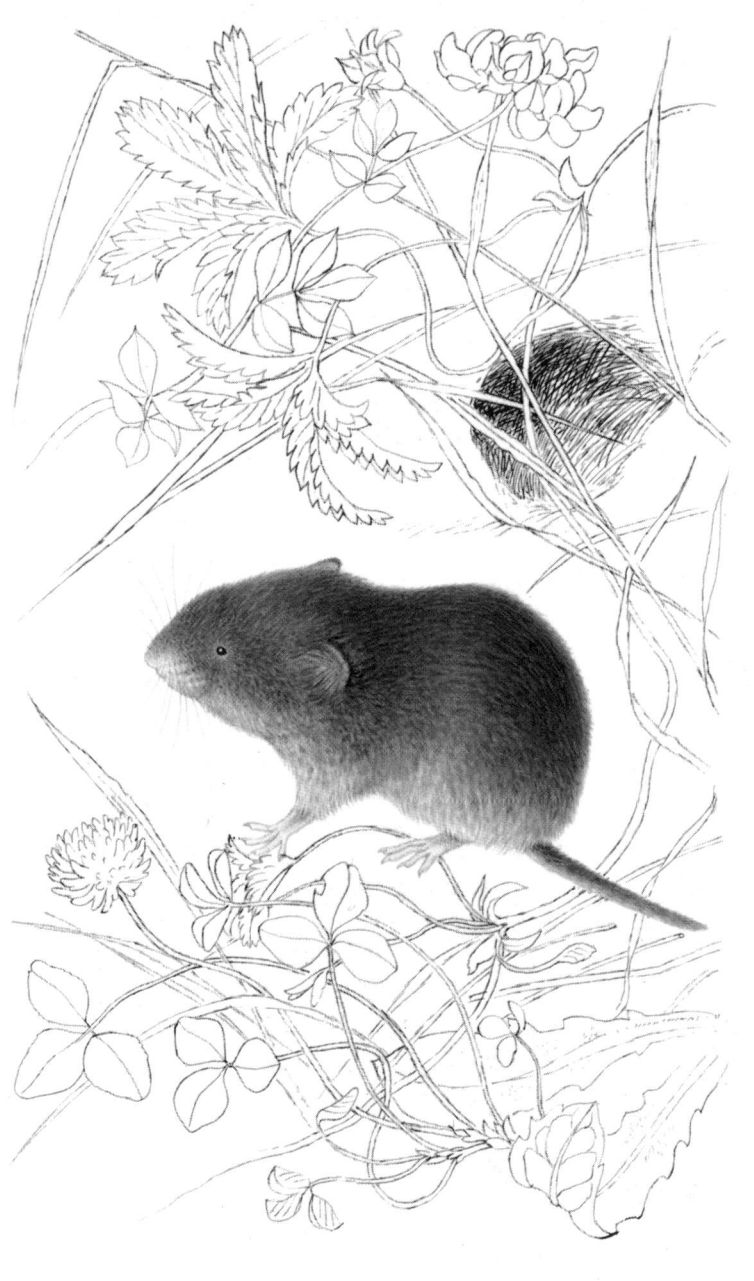

Common Vole or Continental Field Vole
Microtus arvalis
Short-tailed Vole or Field Vole
Microtus agrestis

Microtidae

The Continental field vole is mostly found in cultivated grasslands, such as fields and meadows. It apparently reached Europe with the development of agriculture and, thanks to its adaptability, colonized all suitable habitats from the lowlands up to the mountains. In years of overpopulation, it spreads along paths and railroads to closed forest areas and persists even in wet places and in some alpine meadows. Today it lives over the whole of Europe, except the British Isles, Scandinavia and part of the Mediterranean, but including Orkney and Guernsey. The Continental field vole is a very prolific animal. In favourable conditions, it rapidly increases in numbers, causing enormous damage in fields. It digs its burrows just below the surface of the ground, so that in years of overpopulation the fields, meadows and field boundaries are literally riddled with millions of passages and covered with a thick network of paths. Like all other voles, the Continental field vole mainly eats the green parts of plants, a small amount of seeds and only occasionally insects. It is active during the daytime as well as at night. In winter it builds its nest of dry grass just below the snow cover.

The similar short-tailed vole inhabits wetter and cooler habitats, such as the edges of water, damp meadows, marshes, peat bogs, the margins of mountain streams and forests with undergrowth. It is therefore more likely to be found at higher elevations (up to 1,800 m). It is the common species of northern Europe and England, where it replaces the Continental field vole. The related meadow vole *(Microtus pennsylvanicus)* of North America occurs in similar habitats.

Microtus arvalis:
Body length:
90 — 120 mm.
Tail length:
35 — 40 mm.
Weight:
18 — 40 g.
Litter:
4 — 10 young three to seven times a year.
Life span:
approx. 1 — 1.5 year.

Microtus agrestis:
Body length:
95 — 130 mm.
Tail length:
30 — 47 mm.
Weight:
25 — 55 g.
Litter:
3 — 6 young three to four times a year.
Life span:
approx. 14 months.

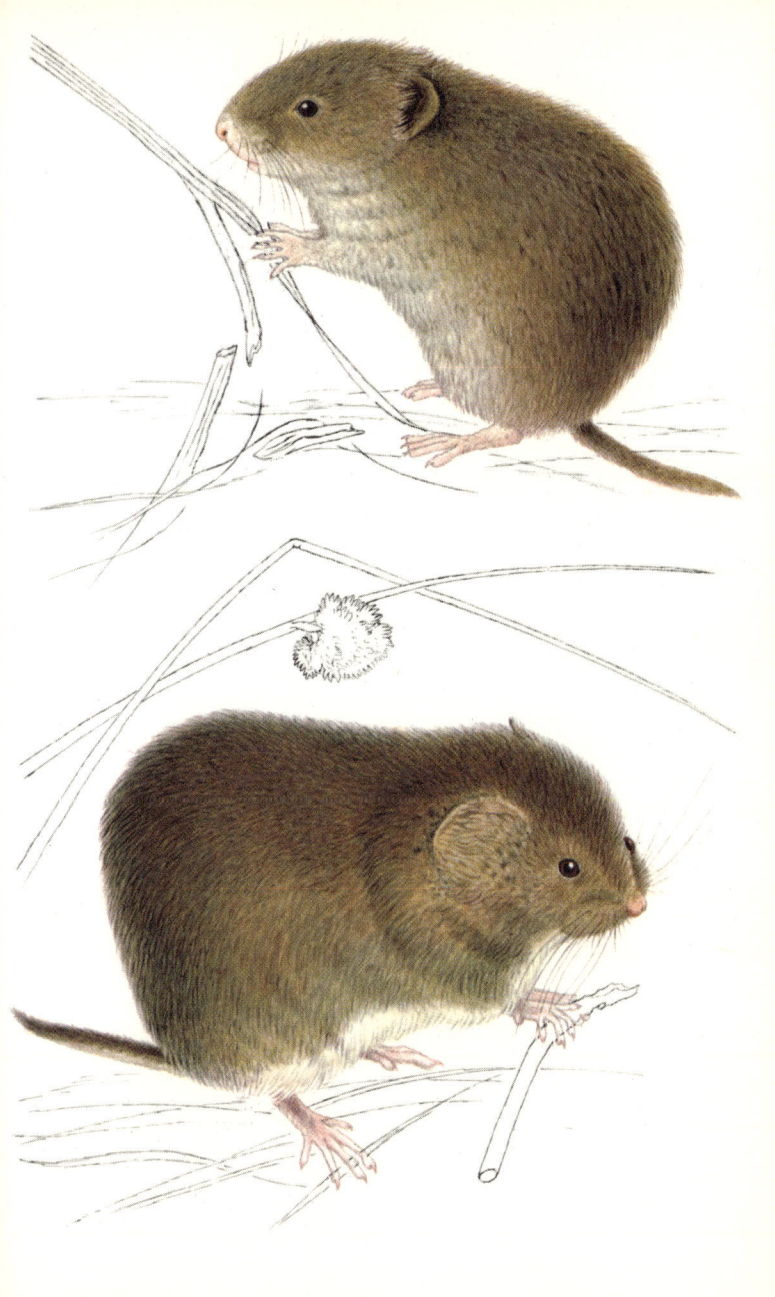

Lesser Mole Rat
Spalax leucodon

This rodent is perhaps the most peculiar of all European mammals, being perfectly adapted to an underground way of life. Its cylindrical body is covered with very short, thick, glossy fur, its eyes are completely overgrown with haired skin and external ears are altogether absent. Its strange large, flat head, with protruding huge incisors, is ornamented with a row of hard, light-coloured whiskers on the cheeks. With the help of its powerful incisors, the mole rat tears, rather than digs, its underground passages, using its forelimbs with their large claws to remove the excavated soil. The passages run not far below the surface of the ground, and only the winter burrows are located at greater depths (up to 2 m). The soil excavated from the passages is thrown up to the surface in the form of huge mounds similar to molehills. This peculiar rodent inhabits the Balkans, with Hungary as the northernmost limit, south-eastern Europe (especially Ukraine), Asia Minor and Transcaucasia. It occurs from the lowlands high up into the mountains (2,400 m) and is to be found in fields, meadows, steppes and on mountain slopes. The lesser mole rat rarely abandons its burrow, feeding mainly on the subterranean parts of plants and occasionally also various invertebrates. For the winter it lays in underground stocks of food, which may sometimes consist of up to 15 kg of sugar beet, potatoes and roots. It does not sleep in winter but its activity is reduced.

The North American pocket gophers (family Geomyidae), although quite unrelated, are very similar to the mole rat in appearance and habits.

Body length:
150—240 mm.
Tail:
reduced.
Weight:
140—220 g.
Litter:
1—4 young
once a year.
Life span:
unknown.

Harvest Mouse

Micromys minutus

In general, mice are not among Man's favourite animals. However, anyone who had the opportunity of watching it closely, could not fail to like the harvest mouse. It is a minute animal, the smallest of all European rodents, and has a rusty yellow coat, pure white underside and large, black eyes. It is active all the time, running quickly about and skilfully climbing blades of grass and plant stems with the help of the prehensile tip of its long tail. The harvest mouse is neither common nor continuous in distribution. Originally it was an inhabitant of densely overgrown marshes and reedbeds, but today it can be found in damp meadows, vegetation by pond margins and wet places in cornfields. For the winter it retires to undisturbed vegetation, haystacks and often also to buildings. Unlike other mice, it builds a neat spherical nest woven of grass, which it hangs among reeds, the stalks of cereals or among other plants, 20 to 80 cm above the ground. In this way it adapts itself to living in places which may often be flooded. The harvest mouse's area of distribution covers the whole of Europe, with the exception of Spain, Ireland, Scandinavia and the southernmost regions of Italy and the Balkans; it also inhabits the whole of Asia as far as Japan and North Vietnam. However, its incidence in this area is confined to limited places where there are suitable conditions. In Britain it is locally abundant, but most often escapes Man's attention in the dense tangle of grasses.

Body length:
82—108 mm.
Tail length:
70—100 mm.
Weight:
5—9 g.
Litter:
3—7 young two or three times a year.
Life span:
1.5 years in the wild, up to 4 years in captivity.

Yellow-necked Mouse
Apodemus flavicollis
Woodmouse or Long-tailed Field Mouse
Apodemus sylvaticus

These woodmice are widely distributed and very abundant in Europe. They both have large dark eyes and prominent ears, and are grey-brown to golden-brown above and buff underneath. The yellow-necked mouse differs from the woodmouse in being considerably larger with brighter colours. The brown of its back is in sharp contrast to its pale belly (unlike the woodmouse) and it has a distinct collar of yellow-brown fur which gives it its name. These are both essentially woodland mice, though the woodmouse can often be found in fields far from trees. They inhabit both coniferous and deciduous woodland, where they feed predominantly on the seeds of the forest trees but also consume fair amounts of insects and other invertebrates. Their nests are underground, the burrows being dug in the soft forest soil. Both species are strictly nocturnal and have surprisingly large home ranges. Woodmice are extremely agile, running and jumping rapidly and climbing with great skill. In the more eastern parts of their range, the two species become more difficult to tell apart and present problems even to the specialist. The distribution of the woodmouse in Europe is virtually complete, and it is only absent from northern Scandinavia. Its range includes almost every habitable island on the continental shelf and also Iceland. The yellow-necked mouse is absent from Ireland, Scotland, the Iberian Peninsula, most of France, the Benelux countries and Italy.

The North American deermice (genus *Peromyscus*) are unrelated, but are exceedingly similar to these mice in appearence and habits.

Apodemus flavicollis:
Body length:
98 — 116 mm.
Tail length:
90 — 127 mm.
Weight:
18 — 25 g.
Litter:
3 — 8 young two to four times a year.
Life span:
in the wild 1.5 — 2 years.

Apodemus sylvaticus:
Body length:
82 — 108 mm.
Tail length:
70 — 100 mm.
Weight:
18 — 25 g.
Litter:
2 — 8 young three to five times a year.
Life span:
in the wild 1.2 — 2 years.

Striped Field Mouse

Apodemus agrarius

In coloration, the striped field mouse resembles both the yellow-necked mouse and the long-tailed field mouse. However, it can easily be distinguished by the conspicuous black stripe running down the middle of its back. Among other small European mammals, only birch mice possess this stripe but they are smaller in size and have a strikingly long tail. The striped field mouse is supposed to have come originally from the East. Its present area of distribution extends from Japan and Korea across Central Asia to southern and central Europe, Rhineland being the western limit of the range. Its distribution throughout this area is only local, and it shows preference for wood margins, bushy places alongside streams, parks and gardens. It is usually found in lowland areas or at moderate altitudes, but it will spread along roads as high as 1,800 m. Its nest is located underground in a maze of burrows which it digs itself. As distinct from other field mice, it is not expert at climbing and jumping. It is active during the daytime as well as at night and lives mainly on vegetable matter. It is a very prolific animal: in some years it overpopulates and invades fields and other places where it is normally completely absent. Its way of life does not otherwise differ from that of other species of field mice.

Body length:
80 — 115 mm.
Tail length:
65 — 92 mm.
Weight:
16 — 35 g.
Litter:
5 — 7 young three to four times a year.
Life span:
1.5 — 2 years, in captivity up to 4 years.

Black or Ship Rat
Rattus rattus
Brown or Norwegian Rat
Rattus norvegicus

The brown rat and the black rat are two of the most notorious enemies of Man who, however, has contributed to their spread all over the world. The brown rat is larger than the black rat and has shorter ears and a shorter tail. It is also more common, at least in Europe, and therefore causes the greater part of the damage ascribed to rats in general. It was originally a native of the marshy areas of north-eastern Asia, whence it has spread to the west both spontaneously and by the help of Man's activities. Its invasion of Europe, which dates back to the seventeenth and eighteenth centuries, was apparently assisted by rail and sea transport. Today it is mainly found in human habitations, especially in the sewers of large towns, slaughterhouses, farms, stables, storehouses and cellars. It is an omnivorous animal and in times when food is scarce it will not hesitate to attack even larger animals than itself. It lives gregariously in families, with a strict social structure. It causes great damage in food stores, and transmits various infectious diseases (e. g. Weil's disease).

The black rat is smaller in size, more slender, and is coloured grey, fawn or black. Originally a native of the tropical regions of the Old World, it has spread all over the world on board ships. It inhabits warm places, attics, wooden buildings and granaries. It has an erratic area of distribution, being found especially round sea ports and water-courses. It used to be more common, and in the Middle Ages was greatly feared because it transmitted plague epidemics.

Rattus rattus:
Body length:
160—235 mm.
Tail length:
190—240 mm.
Weight:
150—250 g.
Litter:
5—10 (13—21) young three to six times a year.
Life span:
2—4, max. 7 years.

Rattus norvegicus:
Body length:
190—270 mm.
Tail length:
130—230 mm.
Weight:
275—500 g.
Litter:
4—12 (sometimes up to 22) young two to three (occasionally four times a year.
Life span:
2—4 years.

House Mouse
Mus musculus

The original home of the house mouse was in the Asiatic and east European steppes, where it lived mainly on the seeds of grasses. With the beginnings of early agriculture, the house mouse became a companion of Man (this kind of association is called commensalism). With the spread of cultivated grassland, it has colonized the whole of Asia and Europe and has also been introduced to other continents. Thanks to Man's activities, its present distribution is completely worldwide. The house mouse is harmful to people living in its immediate vicinity, feeding on their food stores and emitting a disagreeable odour; it can also transmit various infectious diseases. It is hardly possible to count all the kinds of damage it can cause. Today a number of forms of the house mouse are known, differing in appearance and way of life. In western Europe it is represented by a long-tailed, dark form (below), which lives for the whole year close to human habitations and their immediate vicinity. Central and northern Europe is the home of a related, more lightly coloured form (above), which often spends the summer season away from inhabited buildings, usually in fields. The original wild form of the house mouse, which lives in the open throughout the whole year, is still to be found in eastern Austria, Hungary, Ukraine and in the Balkans. Commensal populations of the house mouse usually reproduce the whole year round and their fertility is enormous. They are also immensely adaptable and omnivorous, and they can live and reproduce even in mines and cooling plants.

Body length:
73 — 102 mm.
Tail length:
67 — 95 mm.
Weight:
15 — 28 g.
Litter:
4 — 8 young five to seven times a year.
Life span:
max. 2 years in the wild, up to 5 years in captivity.

Northern Birch Mouse
Sicista betulina

In appearance, the northern birch mouse looks like a small house mouse, but zoologically it belongs to a different rodent family which includes the American jumping mice *(Zapus)* and which is related to the well-known jerboas of the deserts. Its closest relatives inhabit the temperate zone of Asia and North America. The black stripe running down the centre of its head and back is a good distinguishing feature, as well as its extremely long tail which is not possessed by any other European rodent. As distinct from true mice, it has four cheek teeth in the upper jaw (true mice have only three). This strange rodent is distributed in the forests, swamps and meadows of northern and north-eastern Europe, in the temperate regions of Asia, and sporadically also at higher altitudes (700 — 1,850 m) in central Europe. Its occurrence in mountainous regions is obviously a relic of its wide distribution in the past. The northern birch mouse is predominantly a nocturnal mammal and is nowhere very common; thus little is known about its life. In summer it builds a nest of grass and moss, locating it above the ground in dense growths of vegetation; in winter it hibernates curled up in an underground nest. But even during the summer, if the temperature drops suddenly, it will easily fall into a state of torpor. It feeds on buds, flowers, seeds, berries and insects. It can climb well with the help of the twisting tail. Another very similar species, the southern birch mouse *(Sicista subtilis)*, inhabits the steppe areas of south-eastern Europe.

Body length:
50 — 76 mm.
Tail length:
76 — 108 mm.
Weight:
5 — 13 g.
Litter:
2 — 7 young
once a year.
Life span:
up to 3 years.

Common or Crested Porcupine
Hystrix cristata

Hystricidae

Porcupines are inhabitants of the warm regions of Asia and Africa; only one species, the common or crested porcupine, is found in Europe, occurring in southern Italy and Sicily. Many zoologists, however, doubt the origins of the European populations, assuming that it was brought to this area by the Romans. It is not necessary to describe this animal in detail. Everyone can recognize it by its long, black-and-white spines which form the covering of its back and flanks. When attacked the porcupine erects the spines, shakes them and rattles its spiny tail. The spines are attached to the skin relatively loosely and, when provoked, the porcupine will make a sudden sideways lunge at its attacker, followed by an instant withdrawal. Since this appears instantaneous and as it usually results in several spines penetrating the attacker, this has led to the belief that the porcupine can shoot its spines. The common porcupine lives either solitarily or in small family groups in dry, bushy habitats, especially in foothills but also often in the vicinity of human settlements. It hides in rock crevices or digs its own burrows. It is a nocturnal animal and is guided first and foremost by the senses of hearing and smell. It feeds on vegetable matter, mainly the green parts of plants, roots, tubers and field crops, only occasionally adding to its diet some animal food. It searches for food as far as several kilometres from the burrow. The young are born with their eyes open, and are covered in soft spines.

Body length:
570 — 680 mm.
Tail length:
50 — 68 mm.
Weight:
10 — 15 kg.
Litter:
1 — 4 young
once a year.
Life span:
10 — 15 years, in captivity up to 20 years.

Rabbit
Oryctolagus cuniculus

The rabbit is a native of the western Mediterranean (north-west Africa, Spain) and it started its spread throughout Europe about a thousand years ago. It was also introduced successfully into Australia, New Zealand and Chile. Its present area of distribution embraces a large part of Europe, including the British Isles, southern Scandinavia and Poland. Its populations, however, do not reach great numbers due to repeated epidemics of the rabbits' illness — myxomatosis. The wild rabbit shows preference for sandy places in open country, both lowlands and hills, with a dry, warm climate. It can be found on the margins of woods, on bushy, stony slopes, fallow land, pastures, along roadsides and railway embankments, and in gardens and parks on the outskirts of large towns. It seldom occurs at higher altitudes (above 600 m). During the daytime it mostly stays in its underground burrow (up to 3 m deep) and only at dusk, and especially at night, does it come out. It keeps close to its warren and rarely leaves it for a distance farther than 600 m. Rabbits are very prolific animals; they mate from February to July and the female, after a pregnancy of 28—31 days, produces repeated litters which follow quickly one after another. The naked, blind young are born and reared in special short burrows, the entrances of which are blocked by the female. Wild rabbits are popular game animals.

Body length:
350—450 mm.
Tail length:
40—73 mm.
Weight:
1.5—2
(occasionally 3)
kg.
Litter:
4—12 young four
to seven times
a year.
Life span:
max. 10 years.

Brown or European Hare

Lepus europaeus (L. capensis)

Mountain Hare

Lepus timidus

Leporidae

The brown hare is a well-known animal in Europe and is an important game species. It is a native of grasslands and therefore is most abundant in open agricultural land, especially in lowland and hilly country. It is only occasionally found in woodland or in mountains. Its area of distribution comprises all of Europe, with the exception of northern Scandinavia. It is replaced by a similar form in the greater part of the Iberian Peninsula and some of the Mediterranean islands. It also inhabits North Africa and western parts of Asia. The brown hare was introduced into North and South America, Siberia, Australia and New Zealand. Its original area of distribution is expanding continually, particularly in the USSR. Unlike the rabbit, it hides in a shallow open nest, a form, relying upon its camouflage coloration. With the exception of the mating period, it lives alone. The mating time lasts from January to October, and the furred young are born with their eyes open.

The mountain hare is a smaller northern species with shorter ears and thickly furred paws; its coat turns white in winter. It is found in Iceland, Scotland and Ireland, in Scandinavia, northern Poland, northern regions of the USSR, Greenland and Canada. An isolated population, obviously a relic from the Ice Age, lives in the Alps at altitudes of 1,200 — 3,400 m. The American Arctic and snowshoe hares are very similar. Other American hares of the genus *Lepus* are called jack-rabbits.

Lepus europaeus:
Body length:
550 — 650 mm.
Tail length:
75 — 100 mm.
Weight:
3.5 — 5 kg.
Litter:
1 — 5 young three to four times a year.
Life span:
7 — 8 (12.5) years.

Lepus timidus:
Body length:
460 — 610 mm.
Tail length:
40 — 80 mm.
Weight:
2 — 4 kg.
Litter:
2 — 6 young two to three times a year.
Life span:
8 — 9 years.

Lepus timidus:

Wild Boar

Sus scrofa

Wild boars are ancient inhabitants of European woods. In the nineteenth century they were completely exterminated in many places and their numbers did not increase again until the post-war years. These large mammals are only rarely found in the wild. They do not leave their shelters until dusk, and then move very cautiously about, guided by their excellent senses of smell and hearing. During the daytime they usually stay hidden in dense forest undergrowths or in marshy places. Females and young of various ages live in herds the whole year round. Old males live alone. They like to roll in mud and then rub off the dried mud from their skin against tree trunks, thus marking their territories. Wild boars are truly omnivorous animals, consuming anything they can find, from field crops, seeds of trees (e. g. acorns, beechnuts) and roots to insect larvae and carrion; they also hunt fish in shallow pools and root for small rodents in their burrows. Wild boars can cause great damage in fields and meadows, but on the other hand they destroy the larvae of harmful insects in the forests. Their mating time lasts from November to January. The striped young are born in March or April, usually in a shelter lined with grass and moss.

Body length:
110 — 180 cm.
Tail length:
15 — 20 cm.
Weight:
50 — 200 kg.
Litter:
4 — 12 young once (occasionally twice) a year.
Life span:
10 — 12 years.

Red Deer
Cervus elaphus

The red deer is the real king of the woods. It inhabits a vast area from the Iberian Peninsula and North Africa through the whole of temperate Eurasia to North America. Within this territory there are a number of subspecies which differ in very conspicuous features and are sometimes classified as independent species. In Siberia, for instance, there is the large Altai wapiti, also called the maral *(Cervus elaphus sibiricus)*, and in North America the wapiti *(Cervus canadensis)*. In Europe, the red deer is abundant only in the east, especially in the Carpathians and the Alps; isolated populations also live in western Europe, as far north as Scotland and southern Norway. The red deer was originally an inhabitant of the wooded grassland belt; at present it is in most cases found in mixed mountain woods. It is also kept in enclosed preserves. Red deer usually remain hidden during the daytime, coming out to graze at dusk, at midnight and at dawn. Their diet consists mainly of grass and various herbs, young shoots, leaves of trees and shrubs, beechnuts, acorns and field crops. The most interesting period of a red deer's life is the rutting time when the herds break up and the stags, which up to this time have been living a solitary life, collect groups of hinds around themselves. They announce their territory with a low trumpeting call and sometimes also fight against each other. The young are born in May or June after a pregnancy of 8.5 months and are suckled for 3 — 4 months. They become independent after a year.

Body length:
165 — 250 cm.
Tail length:
12 — 15 cm.
Weight:
100 — 350 kg.
Litter:
1 young
once a year.
Life span:
15 — 20 (max. 25)
years.

Sika Deer
Cervus nippon

The small sika deer of eastern Asia is closely re-. lated to our native red deer. In many parts of Europe it is kept as an ornamental deer in parks and enclosed preserves and has also been introduced in the wild. Its original home is the Ussuri region of the USSR, Manchuria, China, Korea and Japan, where there are several subspecies. The sika deer can be easily distinguished from other deer. It is smaller than the fallow deer and its antlers have a maximum of eight, very occasionally ten, points. Consequently, it is not much valued as a game animal. In Europe it has been introduced into Denmark, Germany, England, France, Czechoslovakia and the USSR. It has also been introduced to New Zealand. It is not difficult to keep in captivity; it withstands the winter climate and usually completely loses its original shyness. The males shed their antlers in May or June, and the rutting season begins in the second half of October. The young are born in May or June and are guided by the hind till the end of the winter, with an interruption during the mating period. The sika deer is sometimes mistaken for the axis deer *(Axis axis)*, a native of the open woods of India and Sri Lanka. This species has been introduced into Europe from the beginning of the eighteenth century and can be found in places in Germany, England, Austria, Czechoslovakia, Yugoslavia and elsewhere.

Body length:
110—130 cm.
Tail length:
10—15 cm.
Weight:
25—110 kg.
Litter:
1—2 young
once a year.
Life span:
15—20 years.

Fallow Deer
Dama dama

The fallow deer is a native of the eastern Mediterranean and North Africa, but it was introduced to the whole Mediterranean region and western Europe in ancient times. By the Middle Ages it had reached central Europe, where it was originally kept as an ornament in enclosed preserves; only later was it released into the wild. These introduced herds today comprise the chief stock of wild fallow deer. The original wild populations have in the meantime been almost completely exterminated; their only remnants survive in the woods of southern Iran and adjoining Iraq and are sometimes classified as an independent species, *Dama mesopotamica*. Fallow deer are somewhat smaller than the red deer and differ from them in a number of typical features, mainly their palmated antlers, spotted coloration and a black and white patch on the rump. Completely dark, unspotted individuals are often kept in enclosed preserves. In Europe the fallow deer's favourite habitats are open deciduous lowland forests or mixed, park-like woods. The rutting season falls later than in red deer, usually in October and November. The fallow buck's antlers finish growing in September and are not shed until May. The young are born after a gestation period of eight months. Fallow deer are active during the daytime as well as at night and live in herds. They do not differ markedly from red deer in their way of life, but they show less endurance in running.

Body length:
130 — 160 cm.
Tail length:
16 — 19 cm.
Weight:
45 — 100 kg.
Litter:
1 young
once a year.
Life span:
15 — 20 years.

Reindeer
Rangifer tarandus

The reindeer is the characteristic deer of the north. Numerous subspecies are distributed throughout the tundra and northern taiga of Eurasia, North America and the most of the polar islands. One of the reindeer's adaptations to life in the north is that its two main hooves are broad and outspread. The lateral hooves are also relatively large and are placed lower than in most deer. This modified foot prevents the animal from sinking into snow and muddy soil when it is running. Reindeer antlers are richly branched and have many points. They are worn by both sexes, which is exceptional in the deer family. Especially in the American continent, reindeer are known to undertake long yearly migrations from the tundra to the taiga, and *vice versa*. They are not fastidious animals, and find enough food in lichens, moss, grass, the leaves and twigs of willows, and other polar vegetation. These qualities have lead to the domestication of reindeer in polar regions, and life in the north today is hardly imaginable without these familiar animals. The wild form of reindeer, which used to be very abundant in the north, has been reduced by hunting to such an extent that some subspecies are almost extinct. In Europe small numbers of wild reindeer survive only in northern Norway, Finland and the USSR. Reindeer are expert runners and swimmers, and are active mostly during the daytime. They converge into herds consisting of females, young and young males. The rutting time falls in September and October, and the calves are born in May and June.

Body length:
130 — 220 cm.
Tail length:
7 — 20 cm.
Weight:
60 — 315 kg.
Litter:
1 — 2 young
once a year.
Life span:
12 — 15 years.

Elk or Moose
Alces alces

The elk is an ancient inhabitant of the northern forests of Eurasia and North America. In the Middle Ages it also used to live in the vast forests of central and western Europe, where it later became extinct. Thanks to the protective measures taken during the past few years, elk populations have increased and elks have again spread southwards to the boundaries of their former area and also to the northern tundra. They have already colonized immense areas of the USSR and Poland and continue to spread into Czechoslovakia, Germany and Austria. In the summer, elks live either alone or in families. In winter, after the rutting season, they congregate in small herds numbering 5—10 individuals. Except for the seasonal migrations, which are influenced by the density of population and rutting activity, elks are faithful to a restricted home range. They do not defend their territories in any way. The diet of elks consists of the leaves and twigs of trees and bushes (aspen, alder, willow, poplar), aquatic plants and the shoots of conifers. Their long legs enable them to graze on the leaves of trees; they also eat aquatic plants half submerged in the water. When they are gathering food from the ground, they often bend their forefeet back and support themselves on their wrists. The rutting season lasts from September to November and the antlers are shed between the end of April and the beginning of June. Elks are guided predominantly by their senses of smell and hearing, their vision being very poor. Experiments with the domestication of elks are being carried out in the USSR.

Body length:
250—270 cm.
Tail length:
12—13 cm.
Weight:
250—500 kg.
Litter:
1—2 young
once a year.
Life span:
20—25 years.

Roe Deer

Capreolus capreolus

Originally a forest-dweller, the roe deer has adapted very successfully to life in cultivated land and is now becoming an inhabitant of fields and small woods. It is very abundant in Europe and in some places is among the most popular game animals. Its distribution area, however, is not as continuous nor as extensive as that of the red deer. It ranges from western Europe across the temperate zone of Asia to China. It is absent from Iceland, Ireland, northern Scandinavia, the north of the USSR and the Mediterranean. A different, larger race, with more richly branched antlers, lives in Siberia. The roe deer is active both during the daytime and at night, but comes out to graze mainly after dusk or at dawn. Like the red deer, it grazes on grass and herbs, leaves, shoots, berries and mushrooms. During the summer, it lives either alone or in families, but it congregates into larger herds for the winter. Roe deer usually keep to their territories (measuring about 1 sq km) for the whole year round. The rutting season is in July and August but the young are not born until May or June of the following year; pregnancy is prolonged (latent). The young are born hairy and with open eyes and for the first days of their life they are kept in a shelter, the female visiting them only to suckle them. After a week they join the mother who continues to suckle and guide them for a long time.

Body length:
95 — 135 cm.
Tail length:
2 — 3 cm.
Weight:
15 — 30 kg.
Litter:
1 — 2 young
once a year.
Life span:
10 — 12 (max. 18)
years.

White-tailed or Virginian Deer

Odocoileus virginianus

Cervidae

The white-tailed deer is more closely related to the roe deer than to the red deer. Ecologically it replaces the roe deer on the American continent. In its native American home, the white-tailed deer is widely distributed from Brazil to Canada, being absent only in the westernmost parts of the USA and Canada. It was brought from these areas to Europe, and introduced successfully into southwestern Finland and Czechoslovakia (central Bohemia). In its American homeland the white-tailed deer forms a number of geographical races; those of the northernmost regions reach the size of a red deer, while the southern races resemble a roe deer in size. The smallest form, a native of the Key Island off the coast of Florida, is near to complete extinction (about 30—40 animals are left). The white-tailed deer can hardly be mistaken for any other species of the deer family; its antlers are unusually bent at first outwards and than forwards. The conspicuously long tail with a white underside is raised when the deer is in danger or excited, showing a large white patch beneath the tail. This is a similar mode of communication to that of the red deer. Like the roe deer in Europe, the white-tailed deer in America has become adapted to agricultural country and, consequently, its numbers have been steadily increasing. The European populations rut in November. The males make a strange hissing sound, in contrast to the trumpet-like calls of the red deer. A related species lives in Northern America: the mule deer *(Odocoileus hemionus).*

Body length:
85 — 205 cm.
Tail length:
10 — 35 cm.
Weight:
25 — 200 kg.
Litter:
2 young
once a year.
Life span:
10, in captivity
20 years.

Chamois

Rupicapra rupicapra

Bovidae

The chamois (pronounced *sham-wa*) is one of the few members of the high-mountain European mammal fauna. It inhabits the Pyrenees, the Alps, the Apennines, the Carpathians, the Balkans, Asia Minor and the Caucasus. It has also been successfully introduced into a number of other regions. Zoologically it is classified in the cattle family, as a relative of the sheep and goats. Both sexes have small, black horns, curved at the tip, which are permanent and grow continually. Chamois spend the summer in the alpine mountain belt in meadows, on steep stony slopes and amongst rocks. In winter or in inclement weather they search for the protection of woods. They climb excellently and jump from one rock to another. Chamois are active only during the daytime, slowly roaming their territories. In the morning they usually ascend to higher altitudes, coming down again in the evening. Their senses of hearing and smell are well developed, but their vision is less keen. They live in herds guided by old females; the old males live alone. In most cases chamois keep to their territories, which they mark by the secretions of their scent glands. They graze on mountain plants, and in the winter they also eat buds, moss, lichens and the needles of dwarf pine trees. The rutting season lasts from October to December, during which time the males often fight among themselves. The young are born in June and July. The Rocky Mountain goat, *Oreamnos americanus,* is closely related; although it is remarkably different in appearance, its habits are very similar.

Body length:
110 — 136 cm.
Tail length:
7 — 8 cm.
Weight:
35 — 40 kg.
Litter:
1 young
once a year.
Life span:
15 — 20 (25) years.

Mouflon

Ovis musimon

Bovidae

The mouflon is the only European species of wild sheep. It is a native of the Mediterranean region, where it still lives wild in Sardinia and Corsica. Very closely related forms live in Cyprus and in the mountains of western and Central Asia. In the eighteenth and nineteenth centuries, mouflons were introduced to enclosed preserves in central Europe, where they became acclimatized and escaped into the wild in places. At present, wild mouflon populations are found in Spain, southern France, Germany, Czechoslovakia and elsewhere. The total number of animals in these countries is estimated as 20,000 individuals, which many times exceeds the indigenous wild populations of Sardinia and Corsica. In their original home, mouflons inhabit rocky terrains in the mountains but in central Europe they prefer dry, stony ground. They have well-developed senses of hearing and smell and are rather shy. The females and lambs live in herds led by an old ewe for the whole year; in the rutting time, the rams also join the herd. Mouflons are not very particular animals and adapt easily to most varied habitats. They graze on plants and the twigs of trees and shrubs, and in winter they supplement their diet with various fruits, mosses and lichens. The rutting time lasts from October to the middle of December, and the lambs are born after a gestation period of 5 months. Huge mouflon horns are valued hunters' trophies.

The North American bighorn *(Ovis canadensis)* and Dall sheep *(Ovis dalli)* are very similar.

Body length:
110—130 cm.
Tail length:
5—10 cm.
Weight:
25—50 kg.
Litter:
1 (2) young
once a year.
Life span:
15—20 years.

European Bison or Wisent

Bison bonasus

Bovidae

The wisent is the largest European ungulate and is closely related to the North American bison. The wisent was originally widely distributed from western Europe as far as inner Asia, but constant persecution caused a rapid decrease in its numbers and it gradually disappeared from many areas. In the Middle Ages it was already rare in Europe and the right of hunting it was granted solely to rulers and nobility. Later still, it became almost completely extinct over the whole area of its former occurrence. By the beginning of the twentieth century, all that remained was a small herd of wisents in the Bialowiez, a primeval forest in Poland, and another herd in Caucasus. Both these remaining herds were completely exterminated during the First World War and wisents had to be newly bred from specimens surviving in zoological gardens. Careful breeding resulted in a successful increase of the wisents' total world numbers to about 1,000 animals, some of which were set free; the rest live in enclosed preserves or zoos. The American bison *(Bison bison)* has had, more recently, a similar history. As distinct from bisons, wisents are forest-dwellers and used to live mainly in deciduous or mixed woods where there were grassy clearings. The rutting time is in August and September, and the calves are born in May and June.

The wisent is sometimes wrongly considered to be the ancestor of domestic breeds of cattle. The domestic cattle's ancestral line starts with the aurochs *(Bos primigenius),* which inhabited similar habitats to those of the wisent. This species is now extinct.

Body length:
310—350 cm.
Tail length:
50—60 cm.
Weight:
up to 1,000 kg.
Litter:
1 (2) young
once a year.
Life span:
25—30 years.

MAMMALS IN DANGER (CONSERVATION)

Experience has shown that in many respects mammals are much more adaptable than any other vertebrates, particularly amphibians and reptiles, and therefore are not so directly threatened by the dangers of civilization. This, however, is only true in European and North American conditions, where the fundamental transformation of the original forest-covered landscape into the modern cultivated land took place many centuries ago, and where there is more interest in conservation. Nevertheless, large-scale cultivation of the soil, accompanied by the disappearance of deep forest, has taken its toll of many native species of mammal. It was mostly large, hunted species which were affected. The sixteenth and seventeenth centuries witnessed the extinction of the aurochs *(Bos primigenius)*, the ancestor of the present European domestic cattle. The wisent *(Bison bonasus)*, and a little later its American relative, the bison *(Bison bison)*, only just escaped a similar fate. Another large ruminant, the elk *(Alces alces)*, was forced by Man to leave the forests of western and central Europe and only recently has it started to re-colonize its original habitats, thanks to strict protection measures and artificial introduction in some places. The disappearance of some large beasts of prey, namely the brown bear *(Ursus arctos)*, the wolf *(Canis lupus)* and the lynx *(Lynx lynx)*, from most of Europe seems to be irreversible. Probably the most fortunate mammal in this respect was the wild cat *(Felis silvestris)* which, although eradicated completely in many places, has still maintained itself in certain isolated localities even in central Europe.

Slightly better off today as far as protection is concerned are the beasts of prey of the family Mustelidae. Nevertheless, the otter *(Lutra lutra)* and the pine marten *(Martes martes)* have already been forced out of many regions thanks to systematic persecution by Man, and at the beginning of the twentieth century they both stood on the verge of extinction. This sad fate seems to have been averted, however, and these two animals are now fairly common in some regions. This is generally true of the pine marten, but the otter is still quite rare almost everywhere, although it has recently colonized the surroundings of some rivers where it was completely absent a few years ago. Another mustelid, the European mink *(Lutreola lutreola)*, has disappeared, for reasons unknown, from most of Europe, remaining today only in eastern Europe, Finland and western France. Its disappearance is particularly mysterious in light of the fact that its close relative, the North American mink *(Lutreola vison)*, which is today being bred in large numbers in fur farms, has in many places escaped into the wild and seems to be quite at home in its new, European conditions.

Of the smaller European mammals, the most threatened today are the most useful ones, the bats, as is shown by the alarming decrease in their numbers in Europe in the last few years. This is probably due to the ever-increasing lack of suitable places for hibernation; the old mining galleries are being destroyed and originally quiet natural caves are being opened to the public. One cannot rule out the possibility that the application of insecticides also has had an unfortunate effect on bat populations, as insects form the only food of many of them. The decrease seems to be more rapid in certain warmth-loving species, whose populations have always been rather less numerous along the western boundary of their distribution. For instance, the lesser horseshoe bat *(Rhinolophus hipposideros)*, until recently quite a common European species, has almost completely disappeared from many of its original localities during the last ten years.

Small terrestrial mammals, such as insectivores and rodents, are as yet the least endangered of all mammals. However, this will be true only as long as their natural habitats are preserved. The snow vole *(Microtus nivalis)*, for example, which is confined to the limited areas of Europe's high mountains, will remain unthreatened by extinction until its original habitats are disturbed. This does not mean, of course, that the small mammalian fauna of such areas is not influenced at all by human activity. The present insular occurrence of originally abundant European rodents, such as the short-tailed vole *(Microtus agrestis)* or the pine vole *(Pitymys subterraneus)*, and on the other hand the wide expansion of the grassland-dwelling common vole *(Microtus arvalis)*, have all been indirectly influenced by human activity, i. e. by the transformation of the original forest and damp habitats into cultivated grassland. Such interference with the natural composition of habitats, although resulting in the decrease in numbers of some species, cannot be generally condemned. The numbers of many mammal species which today inhabit the varied types of new, cultivated landscape have, on the contrary, increased considerably in comparison with their original wild state. It also seems that even the recent intensification of agriculture does not seriously threaten these animals as good, natural shelters are still available to them.

If the countryside is to be conserved for the future, at least in its present state, it will probably not be necessary to intensify the protection of the individual species. Besides bats, however, there are several more mammal species which seem to be threatened by extinction, and the fate of these should be made known to the public. The large-scale application of insecticides, as well as ever-increasing road transport, are the main causes of the continued disappearance of the hedgehog. Similarly endangered are dormice, which suffer particularly from the disappearance of suitable habitats and hiding places.

Which then are the most urgent tasks in current animal protection? First of all, to continue the successful conservation programmes for large mammals. These have led in the past few years to the reversal of their decline in numbers and to the re-introduction of some species into their original distribution areas. It is also necessary to maintain in a natural state various types of sufficiently large regions, such as national parks and protected areas, where complete communities of mammals can be conserved and from which they can spread into the neighbouring cultivated landscape. Finally, the most important task is to acquaint the general public with the mammalian fauna of their own country, so that everyone knows the basic conditions of life at least of the most common species. Man should never forget that animals have the same right to their place on Earth as he has.

HOW TO OBSERVE WILD MAMMALS

If you were to judge the diversity and abundance of mammal life in a given region simply according to what you might see during your daily walks, you would probably come to the conclusion that mammals, compared with, say, birds, are not very abundant. This would be a false conclusion, however, as mammals form a rich and important part of Nature. The reason we do not see them very often is that they mostly lead a nocturnal life, they do not usually make much noise or have recognizable calls and they are generally very shy and, thanks to their keen senses, are always able to avoid Man. In daylight we usually only come across a hare or, less frequently, a deer, a squirrel or a vole, depending on that year's animal populations. To encounter a weasel, a marten or a badger is today an unusual experience. This picture, however, changes considerably if you set out for a walk in the evening, early in the morning or even at night, which is the time when most mammals become active, leaving their hiding places in search of food. Only then, and of course only if you make as little noise as possible, will you realize how rich in mammals your local region is. If you are equipped with a good torch, you will be able to see the animals responsible for the many various rustling sounds which fill the night. You may see dormice, scuttling along the branches of a tree, or the long-legged field mice. With a little luck, you may even see a badger.

There are other species of mammal which are quite abundant in the wild and are even active during the day, but which you very rarely see. Among these are the many small rodents and shrews which live either in underground burrows or under the protection of dense undergrowths. The numbers of these small creatures is best revealed by setting live traps for them, which is the method practised by zoologists. Anyone who wants to get to know the complex mammalian fauna of a given region needs to hunt in all possible types of environment, including built-up areas and even human habitations.

During their hunting expeditions and other everyday activities, mammals usually leave behind various tell-tale signs, which reveal their presence and identity almost as clearly as being able to actually see the animals. The most important of these signs are the paw and claw tracks preserved in sand, mud or snow. A sound knowledge of animal tracks has always been amongst the basic prerequisites of the skilled hunter, and it should be a 'must' for every layman who wants to observe Nature and understand wildlife. It is not very difficult to learn to 'read' the tracks of the most common species of mammal. Not only the shape of the tracks, but also the dimensions of the individual footprints, are very important for track-reading. The table on page 166 was drawn to give the basic footprint measurements of some of the most common European species.

How, then, do we proceed to identify the tracks we find? The best-preserved tracks are usually found in summer in dried mud, but the true season for track-reading is the winter, when the ground is covered with snow. Of course, the quality of the tracks always depends on the thickness, freshness and quality of the snow cover. All a layman can usually do on the spot is to photograph or sketch the tracks and to note down

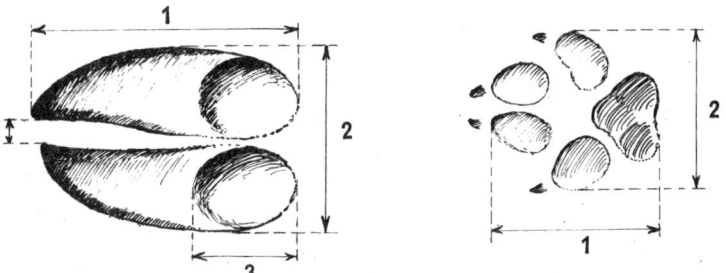

Fig. 4. Footprint measurements:
1 — length, 2 — breadth, 3 — ball, 4 — interdigital distance.

Footprint Measures of the Most Common European Mammals

(dimensions given in cm; F = front foot, H = hind foot)

Species	Footprint length	Footprint width	Pace length
Hedgehog	2.5	1.0	20 — 25
Shrews	0.5 — 1.0	—	3 — 5
Mice and Voles	1.6 — 1.8	—	—
Muskrat	F 3.5 H 7.0	F 3.0 H 5.0	— —
Red squirrel	F 4.0 H 5.0	F 2.0 H 2.5 — 3.5	— —
Beaver	F 3.5 H 15.0	F 4.4 H 10.0	—
Hare	F 5.0 H 6.0	F 3.0 H 3.5	—
Rabbit	H 4.0	H 2.5	—
Lynx	F 6.5 H 7.5	F 5.5 H 6.0	80.0 — 135 (150)
Cat (domestic)	3.0 — 3.5	3.0	30 — 40
Martens	F 3.5 H 4.0	F 3.2 H 3.0 — 4.0	40 — 100
Weasel	1.4	1.0	—
Stoat	F 2.0 H 3.5	F 1.5 H 1.3	30 — 70
Polecat	F 3.0 — 3.5 H 4.0 — 4.5	F 2.5 — 4.0 —	50 — 60
Badger	F 5.0 H 7.0	F 4.0 H 3.5	70 — 80
Otter	F 6.5 — 7.0 H 6.0 — 9.0	6.0	70 — 80
Wolverine	8.0	F 7.0 H 6.5	—
Wolf	F 11.0 H 8.0	F 10.0 H 7.0	80 — 150
Fox	5.0	4.0 — 4.5	30 — 80

Species	Footprint length	Footprint width	Pace length
Bear	F 28.0 H 30.0	F 21.0 H 17.0	150
Wild boar	—	6.0 — 7.0	40
Red deer	♂ 8.0 — 9.0 ♀ 6.0 — 7.0	♂ 6.0 — 7.0 ♀ 4.0 — 5.0	80 — 150
Roe deer	4.5	3.0	60 — 90 (140)
Fallow deer	♂ 8.0 ♀ 5.0 — 6.0	♂ 5.0 ♀ 3.5 — 4.0	—
Reindeer	8.5	9.0 — 10.0	100 — 150
Elk	F 13.0 — 15.0 H 14.0 — 15.5	F 11.0 — 13.0 H 10.5 — 11.0	—
Chamois	6.0	3.5	—
Mouflon	5.5	4.4	—

their basic characteristics. It is only later that the tracks can
be identified by comparing the material gathered with re-
ference books. With some practice, it is possible to keep the
basic types of tracks in one's mind and to read them directly
on the spot without difficulty. The most important prerequi-
site is always to know which features should be observed. The
drawings in handbooks are generally rather idealized pictures
of reality, while the tracks themselves always display a number of variations, depending on the type of terrain, the movement of the animal, the weather conditions and the freshness
of the tracks. The most important features are the tracks'
dimensions. Fig. 4 shows how the tracks should be measured
and which dimensions should be noted. Besides the size of the
track itself, the distance between the individual footprints and
the width of the track are also important.

The characteristic patterns which the tracks form are particularly useful in the identification of species. Here, of course,
one needs to be well aware of the characteristic movements of
the individual species, as gait varies considerably with the
different types and speeds of movement, i. e. walk, run, gallop,
bound, jump, etc. An example is given in Fig. 5A, which shows

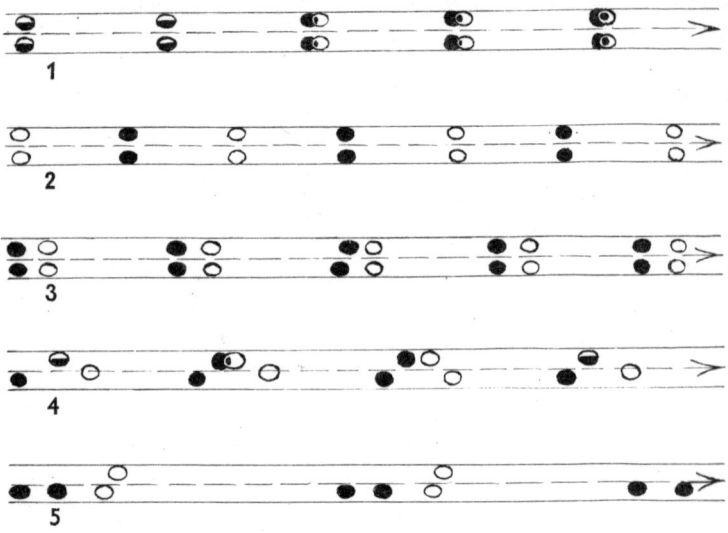

Fig. 5. Various types of footprint patterns:
A — pine marten in slow movement (1, 2) and in fast movement (3, 4, 5)

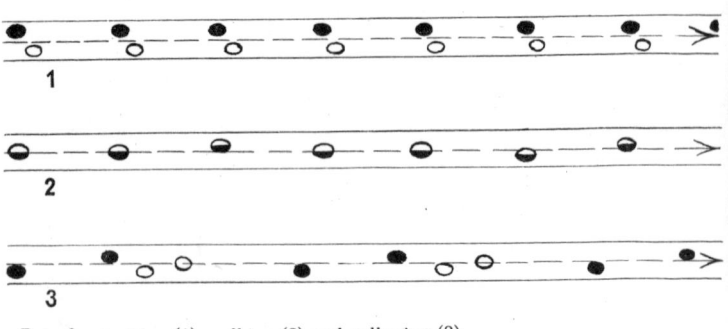

B — fox trotting (1), walking (2) and galloping (3)

various track patterns of the pine marten. The first line shows pairs of footprints, each of which is in fact a double imprint of the respective fore and hind feet. The second line shows

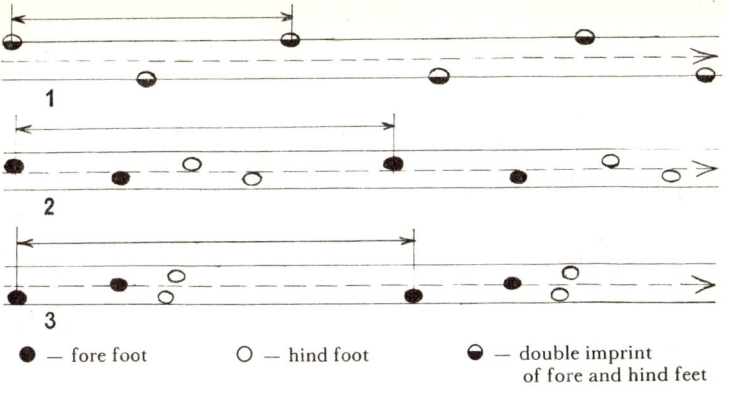

● — fore foot   ○ — hind foot   ◐ — double imprint
of fore and hind feet

C — female deer walking (1), galloping (2) and jumping (3).

regularly alternating fore and hind footprints, which is the
most typical track of a mustelid. These two patterns are the
result of a gentle, slow movement. The following three lines
present quite different track patterns, the common characteristic of which is that the imprints of the hind feet are always
positioned in front of the fore feet. These patterns occur in
many mammals' tracks, as a result of their rapid movement.
The pattern in the last line shows the so-called 'hare's trail',
which is made by the animal's fastest gait and is particularly

Fig. 6. Phases of hare's running and the corresponding footprint pattern.

characteristic of hares. The track pattern of this type of progress is shown in fig. 6. The various track patterns of foxes are also very typical. With this species, sometimes all the footprints lie in an almost straight line, one behind the other at regular intervals, each individual footprint being in fact a double imprint of the fore and hind feet.

In track reading it is also important not to overlook certain minor details such as, for instance, the number of toes, the shape, number and size of the pads on the soles, and also the shape and length of the claws. For example, the otherwise very similar tracks of shrews and some small rodents can be easily distinguished under favourable conditions by the number of toes: shrews have five toes on both their fore and hind feet, whereas the fore feet of rodents are equipped with only four toes. The basic thing the observer has to do is to recognize the imprint of the fore foot, which is vital to each track reading. There are also certain specific features which may help determine some types of tracks. In long-tailed mammals, such as the field mouse and the muskrat the imprint of the tail is also visible in the form of a continuous line between the footprints. Voles sometimes run not only on the surface of the snow, but also dig shallow passages in it. In short, almost every individual track is unique in one way or other and can therefore tell us much about its creator if only we are attentive enough to understand it.

Besides the actual tracks, many mammals leave behind them several other characteristic marks, such as various typical paths and runs, feeding damage on plants and trees, remains of their prey, burrows, scratches in the ground, nests, etc. These not only serve as proof of an animal's presence in the given locality, but may also reveal much about its way of life.

If you follow the tracks of a mammal, sooner or later you usually arrive at a place where it searched for food. The remnants left there reveal not only the composition of the diet, but also the animal's method of obtaining food. Many herbivorous mammals are fond of the bark and young shoots of trees and shrubs. However, it is not always easy to tell with

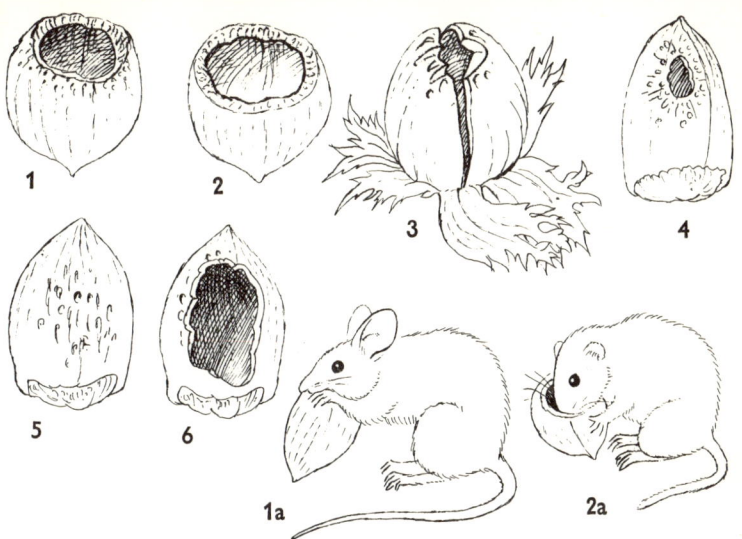

Fig. 7. Hazelnut eaten by 1 — yellow-necked mouse, 2 — bank vole,
3 — squirrel, 4 — tit, 5, 6 — woodpecker.
Typical methods of gnawing a nut shell:
1a — yellow-necked mouse, 2a — bank vole.

certainty which animal has made the scratches on the bark, or
other types of feeding damage. Generally, these might have
been caused by an ungulate, or by a hare or rodent. You
should also always take into consideration various minor details, and above all the location of the damage. If the
scratches are small and located at the base of a tree or shrub,
they were probably made by voles, whereas higher-positioned
scratches indicate the feeding activity of hares or ungulates.
Scratches found in the crowns of trees are most probably the
work of squirrels or red-backed voles, both excellent climbers.
Very characteristic is the feeding damage caused by beavers
or water voles, which are particularly fond of the roots of
fruit trees.

Like many birds, several mammals also feed on the seeds
and fruits of various trees and shrubs; hazelnuts are particu-

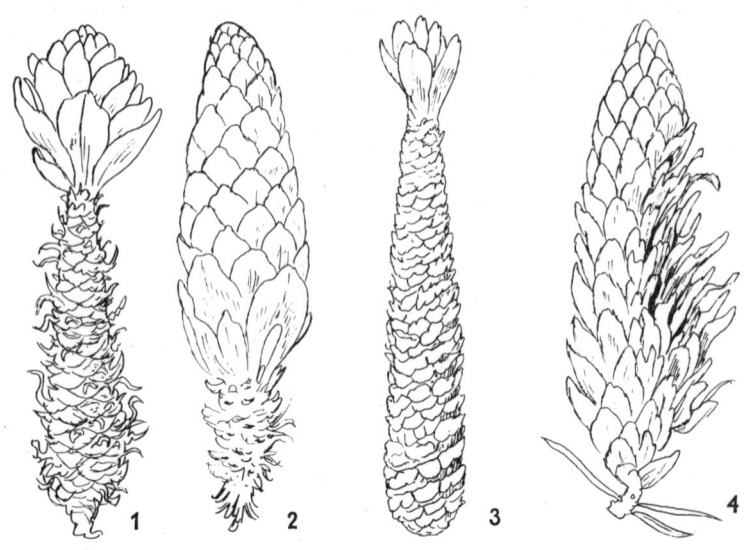

Fig. 8. Cones eaten by 1, 2 — squirrel, 3 — yellow-necked mouse, 4 — woodpecker.

lar favourites with many species. The empty shells can also tell us much about the animal which feasted on them. A squirrel, for instance, will first gnaw a little groove at one end, insert its lower incisors into it and crack the nut open. Field mice and red-backed voles reach the kernel by continuously enlarging a single hole, usually at the blunt end of the shell. Field mice always work on the hole with their lower incisors, while the upper ones hold the shell firm. The outer surface of the shell is therefore always marked with tiny scratches around the hole. Red-backed voles, on the other hand, work differently and the surface of the nut shell remains smooth.

Besides birds, many rodents, particularly squirrels and field mice, are very fond of the seeds of coniferous trees. The remnants of the cones which field mice and voles have feasted upon usually have the scales bitten off very cleanly, whereas those visited by squirrels are torn, rather than bitten, off. Birds such as woodpeckers and crossbills, on the other hand,

Fig. 9. Droppings or scats of some mammals:
1 — hedgehog, 2 — marten, 3 — weasel, 4 — fox, 5 — badger, 6 — hare,
7 — rabbit, 8 — red deer, 9 — roe deer, 10 — mouse, 11 — bat,
12 — long-eared owl's pellets

never tear the scales but only pull the seeds out of the cones. Squirrels, field mice and voles often gnaw the caps of various mushrooms, on which the traces of their tiny teeth are clearly visible.

Very useful sources of information on the mammalian fauna of a given region and the composition of the diet of individual species are the animals' faeces. The droppings of the individual mammal species have their characteristic shape, colour and composition. For instance, the faeces of beasts of prey usually contain remnants of hair and undigested bones, and those of insectivores contain parts of the chitinous wing-covers, whereas the faeces of herbivores contain remnants of plants. The colour of the droppings indicates that not only rodents, but also quite a number of beasts of prey, feed on various berries in the autumn.

Fig. 9 shows the characteristic shape of the faeces of several common mammal species. The droppings of rabbits and hares are quite frequently found. They are almost spherical and those of hares have a diameter of 15—20 mm; the droppings of rabbits are slightly smaller (10 mm) and in certain places are found in large numbers. Very similar are the droppings of deer, but these are a little more elongated in shape and in summer are often glued together in lumps. With the red deer, it is even possible to tell the sex from the faeces; the males produce droppings which are convex at one end and pointed at the other, whereas those of the females have both ends rounded. The size of the faeces can serve as a distinguishing feature in some other deer, namely the reindeer and the elk. The faeces of wild boars are also characteristic, being darkish, irregularly shaped lumps, glued together into larger formations.

The droppings of lesser rodents, mice and voles are cylindrical, with both ends rounded. They consist mostly of plant remains. The faeces of shrews are similar, but they are slightly smaller (2—4 mm), more pointed at both ends and contain insect remains. The droppings of bats have a similar shape and size, but they are rather more porous and are often found in great quantities in places where bats form their colonies.

The faeces of many beasts of prey are also very characteristic. They are usually cylindrical and narrowed into a long point at one end. Their size, composition and various other details can help to distinguish even individual species. The fox, for instance, produces faeces about 8—10 cm long and 2 cm thick, often screw-like and pointed at one end. Very similar are the droppings of badgers, but they are a little more rough on the surface. The faeces of mustelids are usually dark in colour and contain the remains of hair, feathers and fragments of bones. The droppings of hedgehogs very closely resemble those of mustelids: they are also cylindrical, about 3—4 cm long and 8—10 mm thick, dark in colour and glossy on the surface. They are mostly composed of the chitinous remnants of insects. You often come across them on woodland paths in grass, and also in the vicinity of human dwellings.

Another interesting product of mammals' digestion are the so-called pellets—undigested remnants of food from the stomachs of owls and birds of prey, which are regurgitated from time to time and are sometimes found in considerable numbers in lofts, belfries, and under the trees where these birds take their daytime rest. The study of these pellets is of considerable importance to the naturalist, as they comprise not only hair, but also the bones and sometimes even the whole skulls of small mammals, particularly insectivores and lesser rodents, and may therefore not only reveal the composition of the diet of some species of birds, but also offer a relatively accurate account of the small mammalian fauna of a certain locality. This research method has already proved much more accurate than, for instance, trapping or direct observation.

Most mammals do not build any permanent home, but even their seasonal shelters—burrows, lairs, dens, and lodges, which are usually used only during the breeding season—differ in their construction according to the different inhabitants. A surprising number of species hide underground, often digging very complicated burrows with nesting chambers, stores and several entrances. A well-known representative of this group is the mole, whose anatomy is perfectly

175

adapted to an underground life. Its unseen activity is nevertheless soon revealed by the all-too-familiar mole-hills — the piles of earth which have been pushed out from the newly dug corridors. Hardly anybody knows, however, that similar 'mole-hills' are also built by the water vole and, in the south of Europe, by the mole rat. The dimensions and layout of the different 'mole-hills' are characteristic of each species and an expert can easily tell them apart. Relatively deep burrows are dug by the hamster and the ground squirrel and also by the marmot in its mountain habitats. These can be distinguished from the burrows of other rodents by the diameter of the corridors and, above all, by the type of construction, particularly the number and location of nesting and store chambers. As all these species are hibernators, their underground homes have to be built solidly enough to provide a good protection against the winter cold. Some of them also need to house stores of food, mostly grain, for the winter. This is particularly true of the hamster.

The majority of smaller rodents, voles and probably even shrews also dig their own burrows, but it is more difficult for a layman to identify the species from its burrow, as all these animals are approximately of the same size and their way of life is very similar. It is nevertheless possible to detect certain slight differences in construction which point to the owner of the burrow. If it is located in a field and its entrances are connected on the surface by well-trodden paths, then it will probably belong to the field vole. The pine vole, on the other hand, chooses for its home a mountainside meadow, or some part of a forest which is densely overgrown with weeds. The entrances are usually very inconspicuous and well hidden in the surrounding vegetation. Field mice and red-backed voles dig their corridors under the roots of trees or under stones, to provide protection for the nesting chambers. Shrews often use the abandoned burrows of rodents and moles, but if these are not available they will build their own nests, siting them on the ground surface under low vegetation. Many other species which usually dig their own burrows will sometimes, if conditions are favourable, make use of various ready-made shelters.

Field mice, red-backed voles and squirrels are particularly fond of suitable tree cavities, where they not only rear their young, but also make stores of food. The short-tailed vole and the water vole, which inhabit damper localities, build spherical nests of grasses suspended just above the ground on tall vegetation. The same method is practised by the common vole in years of overpopulation.

Wild rabbits, foxes and badgers also hide in burrows, and rear their young there. Rabbits, which often live in colonies, build dense networks of underground passages in clay slopes, with the dens located some 40 — 50 cm under the ground surface. The fox's earth is very easily recognised by the diameter of the entrance. It is usually located in a south-facing sandy bank and, as it is usually used and repaired for a number of years, it often develops into a maze of passages, dens and entrances. Similar in appearance is the badger's set, but it lacks the characteristic 'foxy' smell and is often surrounded by food remnants.

Quite a number of mammals build no permanent home at all, only seeking in time of need occasional shelters in various secluded places. This is the case with the roe deer, red deer and the wild boar. The hare, too, does not dig burrows, but hides in shallow scrapes in the ground which are known as forms. Other species build various surface constructions and nests. Among the most well-known are the beavers' lodges, built of boughs, mud and aquatic plants and consisting of a nesting pit and a space for catching food. The muskrats' constructions are also characteristic, made of heaps of roots and stems of aquatic plants and built in shallow water near the overgrown banks of ponds, and in marshes. Muskrats also build underground burrows in pond banks, with the entrances under the water.

Nests of twigs and leaves built in trees and shrubs are rather exceptional in mammals. Such a nest, called a drey, is constructed by the squirrel: it is an untidy spherical structure made of twigs and tree leaves, with a diameter of 25 — 50 cm. The softly lined nesting chamber is in the centre. Similar, smaller nests of leaves are built by dormice in shrubs and small trees. Perhaps the most peculiar nest belongs to one of the smallest European mammals, the harvest mouse. It is an intricately woven sphere made of grass blades, without a proper entrance — the owner just pushes in — and suspended from grass stalks at a height of 30 — 40 cm above the ground.

A basic knowledge of all the signs and marks left behind by animals is a must for anybody who wants to get acquainted with the mammalian fauna of his home region. Like all other forms of wildlife, mammals do not occur in isolation, but always exist in close relationship to their environment. Each species is found in a certain habitat which for one reason or other conforms to its needs. In some cases, this relationship can be very easily explained, in other cases it is still a mystery to science. Each individual, or a group of individuals, occur within the boundaries of a definite area which is called the home range, or territory, the difference being that a territory is defended but a home range is not. The size of the territory naturally depends on the size of its inhabitant, its food requirements, etc. The vole's home range may measure only some tens of square metres, while those of field mice or shrews are larger. The territories of larger beasts of prey, on the other hand, measure sometimes as much as several tens of square kilometres. The home ranges of herbivores are always smaller than those of preying animals, due to the type of food these two groups of mammals live on. The smallest territories within a species are occupied by young animals and females, the males always staking out much larger hunting areas. The size of a territory also depends on the varying density of population, the type of the habitat and the season of the year.

Anyone who makes a regular study of mammals in their natural habitats soon finds that certain places are visited more often than the rest of the area. Within the given territory, each mammal chooses several favourite spots which it then visits more or less regularly. This is how the characteristic animal paths and 'runs' originate, on which it is possible to come across even the less common and shyer animals. Professional foresters, gamekeepers and experienced hunters are so well acquainted with an area that they are always able to tell beforehand which place will be visited by a certain animal at a given time. With observation and experience, even the layman can acquire this knowledge, a basic requirement for the understanding of wildlife.

BIBLIOGRAPHY

Allen, G. M.: *Bats*. Dover 1962.

Bang, P., P. Dahlström: *Tierspuren*. BLV München, 1973.

Blackmore, M.: *Mammals in Britain*. Collins, 1948.

Bourlière, F.: *The Natural History of Mammals*. Harrap, 1955.

Brink, F. H., van den: *Field guide to the mammals of Britain and Europe*. Boston, 1967.

Carrington, K.: *The mammals*. Time Life International.

Corbet, G. B.: *The identification of British mammals*. British Museum of Nat. Hist., London, 1964.

Corbet, G. B.: *The terrestrial mammals of Western Europe*. G. T. Foulis, London, 1966.

Crowcroft, P.: *The Life of the Shrew*. Reinhardt, London, 1957.

Crowcroft, P.: *Mice all over*. Foulis, London, 1966.

Eisentraut, M.: *Aus dem Leben der Fledermäuse und Flughunde*. Jena, 1957.

Godfrey, G. K., P. Crowcroft: *The Life of the Mole*. Museum Press, London, 1960.

Grassé, P. P.: *Traité de Zoologie. Mammifères,* Vol. 17, 2 parts, Masson et Cie., Paris 1955.

Grzimeks Tierleben. *Säugetiere* Vol. 10 — 12. Kindler Verl., Zürich.

Hall, E. R., K. R. Kelson: *The mammals of North America,* 2 vols., New York, 1959 (The Ronald Press).

Hanzák, J.: *Mammals of Britain and Europe*. Hamlyn, London, 1975.

Harrison-Matthews, L. H.: *British Mammals*, Collins, 1952.

Harrison-Matthews, L. H.: *Animals of Britain* (in separate volumes) The Sunday Times Book Publications, 1962 — 1969.

Harrison-Matthews, L. H.: *The Life of Mammals*. 2 vols.

Hoffmeister, D. F.: *Mammals*. Golden Press, New York.

Hwass, H.: *Mammals of the World*. Methuen, 1961.

Lawrence, M., R. Brown: *Mammals· of Britain*. Blandford Press, London.

Leutscher, A.: *Tracks and Signs of British Animals*. Cleaver-Hume Press, London, 1960.

Morris, D.: *The Mammals*. Hodder and Stoughton, London, 1965.

Mosby, H. S., O. H. Hewitt, 1965: *Wildlife investigations techniques.* The Wildlife Society, Washington, 1965.

Neal, E. G.: *The Badger.* Pelican, 1948.

Petter, F.: *Les Mammifères* in *L'Encyclopédie de la Nature.* Ed. R. Laffont S. A., Paris, 1973.

Siivonen, L.: *The Mammals of Northern Europe.*

Southern, H.: *The Handbook of British Mammals.* Blackwell Scientific Publications, Oxford, 1964.

Stehli, G., P. Brohmer: *The Young Specialist Looks at Animals: Mammals.* Burke, London, 1965.

Young, J. Z.: *The Life of Mammals.* Oxford Univ. Press, 1957.

Vesey-Fitzgerald, B.: *British Bats.* Methuen, London, 1949.

Walker, E.: *Studying our fellow mammals.* The Animal Welfare Institute. New York, 1966.

INDEX
OF COMMON NAMES

INDEX OF LATIN NAMES